Public Regulation of Tumor Banks

Xavier Bioy
Editor

Public Regulation of Tumor Banks

Establishment, Heritage Status, Development and Sharing of Human Biological Samples

Springer

Editor
Xavier Bioy
Toulouse 1 Capitole University
Toulouse, France

Translation from the extended French language edition: "La régulation publique des centres de ressources biologiques" edited by Xavier Bioy, © LEH Édition 2018.

ISBN 978-3-030-08043-3 ISBN 978-3-319-90563-1 (eBook)
https://doi.org/10.1007/978-3-319-90563-1

This Springer imprint is published by the registered company Springer Nature Switzerland AG
The registered company address is: Gewerbestrasse 11, 6330 Cham, Switzerland

Foreword

The research, an account of which we will get over this book, is designed to analyze how the essential resource for research that are human biological elements, mainly tumor tissues, and the personal data associated with them are developed. The treatment and study of cancer requires us to collect and store diseased tissue and thus constitute collections that will be useful to the patient himself, and then possibly to research. For about a decade, these collections have been subject to a legislative framework that protects different interests but also defines the means of a common standard of conservation and circulation. From the beginning, structured collections of tumors, or tumor banks, have figured prominently among biobanks. If we refer to the definition of the INCA, "tumor banks are located in health facilities, where biological samples are recorded, cryopreserved, then according to specific, legally defined conditions and structures, are made available to doctors and scientists for clinical tests or research studies".

In the field of cancer research, they occupy a crucial place in translational research, i.e. research which attempts to use the technical and scientific advances for medical care as quickly as possible. Making samples and personal patients' data available to researchers is a vital link for the progress of care protocols, in particular as it enables us to test biomarkers for prevention and diagnosis.

That is why the topic of this research and this conference focuses on studying, in human and social sciences, the formidable question of the "development" of collections and, concretely their regulation or their governance. Basically, how can be decided nowadays and how could be decided in the future the constitution, the funding, the allocation of resources to a particular research program?

The interests at stake are still incorrectly identified but there are already a great number of them and your conference will enable this identification. The interests of researchers come first but they are already diverse, depending on the purposes of research, but also on the integrated nature or not of the collection to the research program.

But the interests of public and private funders are no less real. The INCA as well as some teaching hospitals have thus perceived the importance of a public policy for these collections, but the health insurance, which funds much of the process, has only done so to a lesser extent. The Inserm has implemented for several years an extensive program of structuring which is beyond the scope of oncology, but which integrates it. Health players are really receptive to research like yours around the new job of biobanker, which tends to move away from the only organ or pathology expertise.

The essential contribution of this research resides in the legal, economic and social study of biobanks in oncology. Indeed, although there are some legislative or regulatory elements, they are far from covering all the important or specific aspects. In particular, the elements of a public policy of development are still pending. The doctors and engineers here will insist on quality requirements and technical standards, but there is also, as you know, the issue of the collections use.

There are many ways to help with the creation of these resources and with making them available for research. All these ways require a clarification of roles and jurisdictions. I have seen that the book will lead to reflect on the constitution of public or private "patrimonies" and to clarify the legal status of the contracts that secure the circulation of collections. It is both necessary and difficult for lawyers to analyze these aspects precisely and they should do so along with doctors and with the managers of these collections to assess practices and guide them.

It is also remarkable that this research has created a group that includes economists, in a prestigious economics school which already analyzes health economics. A fine apprehension of costs and of the market around biobanks, and specifically tumor banks, seems indispensable to know where public authorities can help.

I see that the book makes room for a sociological perspective. It is a great thing because you cannot ignore the representations that society conveys when it comes to cancer and research, the human body, and "human resources".

It seems to me that this book raises the right questions. Those questions are considered from the outset to define the legal and economic nature of collections which interest researchers. I guess it is a challenge in itself to name what already exists.

Next, you will get to the core of the issue, from my point of view: why develop collections and offer broad access to them? Especially through the sound framework of European relations and international cooperation activities? The issue of networking is central to all research, but this is especially true for cancer which requires the collection of a huge amount of information on its determining factors.

Finally, the reader can provide what scientists expect: leads and answers on the legal and political means of this development and of this regulation to benefit the greater good. The book intends to carefully consider ways towards a pooling of resources/knowledge. This issue of 'pooling', regarding health, should not worry anyone. Of course, the cost of new therapies is such that drug producing laboratories must protect their work and their resources. But it is also politically important to encourage, as much as possible, collectors to share their resources for an optimum collective distribution, in the interest of all.

Finally, I am very pleased to see that the Toulouse site has also seized the opportunity of the presence of Oncopole. Indeed, the 2003 project, which has now been completed, combining public and private research with all types of care structures, will have an influence. Researchers from other structures, such as Université Toulouse 1 Capitole, will now work with them more, because cancer is also a matter for human sciences. Toulouse becomes year after year a key location of the fight against the disease.

Deputy Secretary General of the
United Nations, New York, NY, USA

Philippe Douste-Blazy

The original version of this book was revised. An erratum to this book can be found at https://doi.org/10.1007/978-3-319-90563-1_19.

Contents

List of Abbreviations

BRIF	Bioresource Research Impact Factor
CAD	Collecteur analyseur des données (Data analyzer collector)
CCQ	Civil Code of Québec
CNRS	Centre national de la recherche scientifique (France) Scientific Research National Center
CPP	Comité de protection des personnes (France) Committee for the Protection of Persons
CPPP	Contractual public-private partnerships
CRB	Centers of biological resources
CSP	Code de la santé publique (France) Public health code
EPIC	European Prospective Investigation into Cancer and Nutrition
ERIC	European research infrastructure consortium) in the area of biobanks and Biomolecular resources, « ERIC-BBMRI
ELSI	The Ethical, Legal and Social Issues Service
ESFRI	European strategy forum for research infrastructures
FAO	Food and Agriculture Organization
GeCIP	Genomics England Clinical Interpretation Partnership
HIPAA	Health Insurance Portability and Accountability Act (1996) US
hSERN	Human Sample Exchange Regulation Navigator
IBiSA	Biology, Health, and Agronomy Infrastructure
IGAS	General Inspectorate of Social Affairs (France)
INCa	French national cancer institute
Inserm	Institut national de la santé et de la recherche médicale (France) National Institute of Health and Medical Research
IRB	Institutional review board
JTIs	Joint Technology Initiative
MTA	Material Transfer Agreement
mBRCs	Microbiological resource centers
OMC	World organization of trade

RCS	Registre commercial des sociétés (Commercial Register of Companies)
RTD	Research, Technological Development
TFEU	Treaty on the Functioning of the European Union

List of Editor and Contributors

Florence Bellivier University of Paris Nanterre, Paris, France

Xavier Bioy University Toulouse 1 Capitole, Maurice Hauriou Institute, Toulouse, France

Tumor Banks Project, Toulouse, France

Catherine Bobtcheff French National Center for Scientific Research (CNRS), Toulouse School of Economics and Toulouse Capitole University, Toulouse, France

Anne Brouchet Biological Resource Centre (BRC) Cancer, Toulouse University Hospital, Toulouse, France

Anne Cambon-Thomsen CNRS, UMR/INSERM 1027 Université Paul Sabatier, Toulouse, France

Marie-Xavière Catto University Paris I Panthéon Sorbonne, Paris, France

Thibaut Coussens-Barre University Toulouse 1 Capitole, Maurice Hauriou Institute, Toulouse, France

Emilie Debaets University Toulouse 1 Capitole, Maurice Hauriou Institute, Toulouse, France

Nathalie De Grove-Valdeyron University Toulouse 1 Capitole, Institut de recherche en droit européen international et comparé (IRDEIC), Toulouse, France

Anne-Marie Duguet UMR/INSERM 1027, University Paul Sabatier, Toulouse, France

Pierre Égéa University Toulouse 1 Capitole, Maurice Hauriou Institute, Toulouse, France

Carole Haritchabalet University of Pau and Pays de l'Adour (CATT), Pau, France

Toulouse School of Economics, Toulouse, France

Paul Hofman Laboratory of Clinical and Experimental Pathology, Nice Hospital Biobank (BB-0033-00025), University Hospital Federation OncoAge, Pasteur Hospital, Nice, France

Laurent Izac University Toulouse 1 Capitole, IEJUC, Toulouse, France

Yann Joly Centre of Genomics and Policy, Montreal, QC, Canada

McGill University, Montreal, QC, Canada

Laurence Mabile BRIF (Bioresource Research Impact Factor), UMR/INSERM 1027, University Paul Sabatier, Toulouse, France

Fabien Milanovic Sup'Biotech Paris, Paris, France

Christine Noiville French National Center for Scientific Research, UMR 8103, Paris, France

Sophie Paricard Institut National Universitaire Champollion, Albi, France

Institute in Private Law, University Toulouse 1 Capitole, Toulouse, France

Sophie Peries Biological Resource Centre (BRC) Cancer, Toulouse University Hospital, Toulouse, France

Anna Pigeon University Toulouse 1 Capitole, Maurice Hauriou Institute, Toulouse, France

Tumor Banks Project, Toulouse, France

Miriam Pinkesz Centre of Genomics and Policy, Montreal, QC, Canada

McGill University, Montreal, QC, Canada

Emmanuelle Rial-Sebbag DR Inserm, UMR 1027 Inserm, Université de Toulouse, Toulouse, France

Benoît Schmaltz University Toulouse 1 Capitole, Maurice Hauriou Institute, Toulouse, France

Part I
Tumor Banks or the Valorization of a Complex Object

Chapter 1
Is Putting Cancer in Banks a Patrimony To Be Developed?

Xavier Bioy

To Philippe Duffau,

From up above, let him make sure that this work advances, even very little, the conditions of cancer research.

This book brings together the contributions of researchers from all walks of life around a transfiguration of cancer. Once mutant cells are removed from the human body, they become an atypical and fascinating object; they literally embody the evil we lock up to make it talk. Talking about it, talking about its host and making cancer talk is the mission and challenge of doctors who cut and preserve, observe and analyze. But this is also the challenge of those who gather, evaluate and make available to researchers and increasingly digitize and circulate patients' data. This rare and demanding know-how is priceless, but it is an activity of resource development that has not revealed all its secrets yet.

Certainly, a lot has been written about the techniques, objectives and interest of human biological collections, particularly in the field of cancer. Journals are devoted exclusively to them; international networks of researchers are organizing around them. But for the time being, they focus mainly on the medical and scientific part or on the ethical and legal aspects that relate to the collection of body elements (consent and non-objection) or medical and biological associated data. Some work focuses on the question of governance of the collections. But, for now, few publications envisage the issue from the standpoint of the creation of general patrimonies and public policies for the development of what constitutes a resource. Its conservation, as onerous as necessary, occupies research organizations as well as hospitals. It was undoubtedly necessary, to envisage the banks as objects to be developed as patrimonies, to have a French outlook, familiar as we are with the intervention of public

X. Bioy (✉)
University Toulouse 1 Capitole, Maurice Hauriou Institute, Toulouse, France

Tumor Banks Project, Toulouse, France
e-mail: Xavier.bioy@ut-capitole.fr

X. Bioy (ed.), *Public Regulation of Tumor Banks*,
https://doi.org/10.1007/978-3-319-90563-1_1

power and with the theory of public and private patrimonies, to see in a collection of tumors an object to be developed by public health institutions and worthy of the interest of jurists, economists and sociologists.

This publication brings together the material finally obtained following a research project initiated in 2013, in Toulouse, but actually underway since 2015, since an international jury gave their go-ahead to its financing by the Research National Agency through the French projects of excellence initiatives (IDEX-UNITI). This is certainly because of its multidisciplinary approach and its integration of the players of the biobank system in the scientific reflection of social sciences and even more in the context of the Toulouse Oncopole which already innovates remarkably with its mode of operation and its ambitions.

This project, beautifully called TUBA for a less pleasant object (Tumor banks), aims to analyze the current situation of human biological resource banks in cancer research, but also to measure the public interest and the stakes of the private sector to see why and how public power could intervene in this system, which is only partially framed today, politically and legally. The samples and collections of tumors circulate, sometimes a little, sometimes a lot, sometimes for "in house" scientific purpose only, sometimes in the perspective of making them available to third parties. What should be their economic model? What are today's practices? What can be done? What should be done? What legal framework should be used in that respect?

These seemingly simple issues, which are first of all to do with public health policies and economic analysis, actually raise considerable legal problems. They also require consideration from a societal point of view. They mainly imply taking stock of the practices and objectives of research professionals, as well as to become aware of the social representations of these yet little-known objects.

The interest of this field of study, under a perhaps too precise heading (public regulation of Biological Resource Centers in oncology), actually raises a significant number of themes and questions that have first fascinated the jurist that I am before gradually federating the group. It is our aim to complete the specialized and accustomed eye of biobank professionals by that of the law experts of the faculty.

1.1 Purpose of the Project

Biobanks, as Mr. Douste-Blazy has quite rightly mentioned, will be crucial places in the development of biotechnology and more generally healthy economy. It is a subject which is already extensively studied, and Toulouse is already a recognized site of studies thanks to, in particular, the team "Trajectoires d'innovations en santé" from Inserm and their partnership with the BBMRI-ERIC network, but also the teams from Oncopole. Their regulation is thus widely known, but the research carried out here is justified in relation to several stakes:

At a first level, that of professional expertise: indeed, collectors, who are also, for the time being, necessarily researchers, often question the legal nature of their object and the reality of the limits to be respected in the build-up and use of collections. They also question their double hats as researchers and "biobankers".

At a second level, that of legal science: jurists, on all sides, are asked to question the subject itself—aliquots, samples, data, collections and know-how, as well as the existing categories of law which lead to diversify the applicable rules. Different objectives animate these rules which are sometimes contradictory, and this is a state of affairs that has not been specifically studied. Different points of view on each of these points are still commonplace within the legal doctrine: can samples be appropriated? What about collections? What about personal data? Do collectors in a public institution have rights on "their" collection? The diversity of objectives, practices and professional viewpoints was also not examined from all angles and still needed to be. Biobank regulation, especially in cancer, where institutions and funding are found to be more numerous, presents itself as a patchwork: it comprises the regulation of the biological object itself (its collection, conservation, making it available, its re-use, etc.), the regulation of the associated patient data (their collection, their securing, their links to biological samples, their transfer, their anonymization etc.), the regulation of the institutions that house and manage the collections (their legal entity, their status as players in the public service, their prerogatives of public power, the nature of their patrimony, etc.), then the regulation of "know-how" (that of the creation of public or private officials who create the collections according to a logic of value added, the possible intellectual property attached thereto, etc.). And then the collection as a whole and all of the collections do not form exactly the same utility and therefore do not depend on the same regulations. It is this complexity, often obscured by practical analyzes which are not subject to litigation, which should be examined.

And then, beyond the commonalities of all biobanks, also comes the specificity of the field of cancer and the institutions peculiar to this field: the Inca at a national level, the regional Cancéropoles and the Toulouse Oncopole. The particular organization of France regarding the Cancer plan and the high impact of public intervention leads to the questioning of the effectiveness of regulation between public power and the market. "Cancerous tumors" are original as "malignant" objects, since they have little patrimonial interest for the patient and thereby allow one to answer the double question of sampling and consent more easily than in other contexts. Legally, they are, nonetheless, elements like others, which gives a general scope to our study.

1.2 Project Method

The method we have chosen has shown great strength by its flexibility. Starting from a hard core of interested researchers, and sometimes already specialists, the group gradually widened as new issues or difficulties arose. To paraphrase Corneille, setting out we were about fifteen people from Toulouse, but, by a speedy reinforcement we saw ourselves forty on arrival at the conference. A long period of "socialization" has been necessary for everyone to take ownership of the object, although it did not cease to get increasingly complex.

We have had countless discussion meetings and trips to the collection premises. We have sometimes had long discussions before realizing that we were not all talking about the same things whilst using the same words: for example, why does the term tumor bank concern only the solids and not the liquids that biologists also store? Why does the tumor quickly become "something else" rather than a biological element? Why does each type of collection have different uses?

Jurists thought that medical terminology was a base on which they could end their interminable legal qualification quarrels. But nay! It has been necessary to define all the terminology, as we were making progress, when we realized our misunderstandings. More importantly, we had to bring a number of jurists, who were knowledgeable on this or that general area of law, to import the elements of the human body, their associated data and the added value of the collector in their concerns and then to take an interest in this object as off-putting as exotic to us, often under the sardonic or stunned eye of fellow jurists of faculties.

Tumors, their infrastructures, and their managers have thus become little by little objects of civil law, of health law, of human rights, of commercial law, of public and private patrimony law, of micro-economics and of sociology of health. They have volume and facets!

After bibliographical research, we conducted a questionnaire survey of French tumor banks to collect various practical data concerning, inter alia, quantities, tariffs, financing, but also evaluative data concerning the legal or economic difficulties as well as the different development policies. The response rate was hardly satisfactory to produce the desired statistical effects. But, nevertheless, it brought interesting testimonies of a non-negligible form of disparity in practices.

We then proceeded to organize 2 days of study bringing together only the contributors and focusing on the multidisciplinary identification of the object on the first day, and on the second day on perspectives and means of development of the collections and their management. Finally, we had a review conference that has greatly contributed to this publication.

1.3 Architecture of the Publication

This leads to a gradual project comprising three stages: one being the definition of the complex object that is a tumor bank, the second relating to its development according to the different viewpoints, the third looking at the means of this development from the most classic to the most forward-looking. This set of contributions, which far exceeds the case of oncology, therefore progresses from situational analysis of the object to the identification of the stakes and claims of the various players in the sector, each having their vision and ambitions for the collections. Finally, this book proposes innovative approaches to development for research: mostly incentive schemes, which largely correspond to what is done today and could be systematized, up to more coercive processes that could reflect a more radical public political will and which correspond to what French public law has been able to do for a long time (while adapting to today's prevailing economic logics, development being an obligation henceforth).

Acknowledgments The project coordinator sincerely thanks those who contributed to its success: Philippe Douste-Blazy, president of the Fondation Toulouse Cancer Santé, under the patronage of whom the conference was held; Anne Cambon-Thommsen and Emmanuelle Rias-Sebbag for their expertise, the Institut Universitaire du Cancer of Toulouse and the Research Directorate of the Centre Hospitalier Universitaire of Toulouse which have provided their support and expertise; the services of the Université Toulouse 1 Capitole and in particular the university's Research Management Service and the budget-bearing Institut Maurice Hauriou; Damien Fallon for the illustrative image of the poster that gives a kaleidoscope to our object by bringing together brilliantly and aesthetically the theme of the collection, jurists, scholars and savants. But I thank especially Anna Pigeon, who has supported every day for 3 years all the aspects of this research with a patience and an intelligence of situations and people which deserve praise. Finally, I thank the contributors, fellow jurists, economists, sociologists and doctors who have agreed to take part in this strange adventure.

Chapter 2
Biological Samples and Their Sampling Degrees

Thibaut Coussens-Barre

The legal apprehension of biological samples collections has long been based on a holistic approach, considering the collection as an object of study *ut universi* subject to a proper legal framework (Noiville 1997; Bioy 2010; Rial-Sebbag 2008).

Our approach here will be to consider biological samples collections *ut singuli*, through the prism of the elements that make them up. Whether they be biological samples, associated personal data, the know-how of professionals who manipulate these samples or even databases that allow the processing and storage of all this information, this decomposition in several elements will enable each of them to enlighten their legal status and to better grasp their outlines.

The first of them, the immediate element that will be the subject of this study, is biological sample. As a basic support of scientific research, they have never been defined by law. On the contrary, this concept coexists alongside other denominations such as biological sampling, biological resource, element and product of the human body, of human thing (Bayer 2003), of biological matter[1] or biological material.[2] Although close, these expressions are not synonymous with it and generate some confusion for legal experts in search of readability.

If reference is made to Article L. 1243-3 of the Public Health Code, "*collections of human biological samples shall designate the gathering, for scientific purposes, of biological specimens collected on a group of persons identified and selected according to the clinical or biological characteristics of one or more members of the group, as well as derivatives of these samples*". According to this definition, a biological sample would therefore be a biological element which has been the

This chapter was written in collaboration with **Pr. Philippe Rochaix**, pathologist, deputy head of the Department of Pathology, IUCT Oncopole, Toulouse, France.

[1]Art. L. 611-10 of the French Intellectual Property Code.

[2]Art. L. 611-10 of the French Intellectual Property Code.

T. Coussens-Barre (✉)
University Toulouse 1 Capitole, Maurice Hauriou Institute, Toulouse, France

X. Bioy (ed.), *Public Regulation of Tumor Banks*,
https://doi.org/10.1007/978-3-319-90563-1_2

subject of a selection procedure before being placed in a collection constituted for a scientific purpose, and from which can be extracted derivatives. These derivatives are not explicitly mentioned in law.

As such a definition only displace our initial qualification problem, then the question arises as to what biological element is. Pathologists can provide a technical definition, but legal experts will have to settle for a material approach by reconciling biological element to existing legal categories, namely that of elements and products of the human body.[3] This category is broken down into a series of subsets which, by voluntarily broad labels (human blood, organs, tissues, cells and human body products), allows one to encompass all biological specimens that may be worked on and placed within collections (Binet 2017).

As we can see, biological samples would be derived from biological sampling, which is understood as an element or product of the human body obtained according to the modalities determined by the public health code.

The postulate defended in this contribution is as follows: biological samples are only the "finished products" of a wider sampling process which spreads over several stages: from the patient's reception, through the technical operation of collection, the handling of the collected element and the work on this sample, to end with all the acts of provision on this collected material.

At each stage of this process, the actions of professionals would help to change the legal condition of the subject matter, so that its name would be changed not only technically, but also legally. Biological elements would result in biological samples. From being a thing, it would become a good.

Based on the different stages of the "life of a biological sample", we shall endeavor to explain the technical conditions of its treatment and the legal consequences of these activities at each stage.

2.1 From a Full Body to a Biological Element

Since biological samples are initially contained in the full body of a person, access to this resource is only possible in accordance with procedures strictly framed by law (Sect. 2.1.1). On the other hand, professionals must carry out a technical act of collection to bring biological sampling to the legal scene, the former being considered as a common law thing (Sect. 2.1.2).

[3]Category introduced by Act No. 94-654 of 29th July 1994 on the donation and use of elements and products of the human body, on medical assistance in procreation and prenatal diagnosis. This act will amend the nomenclature of the Public Health Code by inserting part I, second book on the "donation and use of elements and products of the human body", Art. L. 1211-1 to L. 1274-3 CSP. The regulations are contained in Art. R. 1211-1 to R. 1261-9 CSP.

2.1.1 The Act of Detachment Circumscribed by the Respect of Human Body

As access to the human body is not free and is broadly framed since the first bioethics laws of 1994,[4] professionals wishing to obtain biological elements for a therapeutic or scientific purpose must respect the cardinal principles presented in the Civil Code and the public Health Code.[5] Although this fragmentation of sources may seem detrimental to the coherence of the legal framework of the human body, these rules are at least complementary.[6]

The inviolability of human body[7] implies, at the end of a preliminary procedure of informing the patient, obtaining consent to any collection of elements or sampling of products of the human body. The act of consent, revocable at any time, shall not be construed as an act of disposition of the person on his own body but more as an authorization made to others to impair his physical integrity by an act of collection of elements or products of the human body. Health professionals are authorized, through consent, to perform a medical act on the body of others. Once in possession of biological elements, they shall dispose of them in accordance with the terms laid down in the consent form and according to the rules of law. The person at the origin of consent, ""*the one who donates" the element does not dispose of it as a thing: they do not dispose of the element, they dispose of their body by allowing the collection for scientific or therapeutic use*" (Prieur 1998). The biological element thus acquires a legal autonomy; a distinct status from the person and the body from which it originates.

At the same time, the collection process will be conditioned by other principles such as its free nature,[8] an advertising ban, or anonymity.[9] Finally, professionals who are considering initiating an activity requiring the preservation of elements or products of the human body will also have to comply with the administrative procedures for declaring and/or authorizing the Articles L. 1243-2 *et sequentia* of the Public Health Code.

---

[4]Three statutes adopted in July 1994 establish the terms generally used since bioethics laws: (1) Act No. 94-548 of 1st July 1994 on the processing of personal data for research in the field of health and amending Law No. 78-17 of 6th January 1978 on information technology, files and freedoms; (2) Act No. 94-653 of 29th July 1994 on respect for the human body; (3) Law No. 94-654 of 29th July 1994 on the donation and use of elements and products of the human body, on medical assistance for procreation and prenatal diagnosis.

[5]The Civil Code dedicates a chapter on the issue, it is entitled « Du respect du corps humain », while the public Health Code covers the issue in two specific books « Protection des personnes en matière de santé » and « Don et utilisation des éléments et produits du corps humain ».

[6]Since the Civil Code focuses more on the protection of the full body and the person being attached to it, the code of Public health organizes the technical conditions of sampling as well as the modalities for the use and circulation of elements and products of the human body.

[7]Art. 16-1 of the Civil Code.

[8]Art. L. 1211-4 CSP.

[9]Art. L. 1211-5 CSP.

The collection can only be carried out under certain conditions and for a purpose provided for by the law. The public health Code thus distinguishes according to whether sampling is performed on a living or deceased person and according to whether it is pursuing a therapeutic or scientific purpose. Practitioners are then free to choose the technique used for the collection, although practices tend to be largely standardized under the influence of national[10] and international institutions.[11]

2.1.2 *The Act of Detachment Materializes the Legal Existence of Biological Samples*

The material act of detachment consists in subtracting from a full body—living or deceased—biological elements intended to be used for a therapeutic and/or scientific purpose. The transition from one purpose to another is possible through requalification procedures.[12]

As a sine qua non condition of access to the raw material used in research, the act of detachment induces several legal consequences.

First, the detachment reveals the biological element to the legal sphere. This very existence, independent from the human body and the legal person, requires special supervision. The legislator has considered this reality by introducing, as early as 1994, the category of elements and products of the human body.

The detachment has the effect of putting an end to the fiction of personalistic authors according to whom the body and the person merge to form just one legal entity (Carbonnier 2000; Cornu 1997). Being no longer assimilated to the person and in accordance with a firmly established *summa divisio* "*the elements and products of the body which are collected from it become, from that moment, things of common law, except when they are integrated into the body of another person* " (Zenati-Castaing and Revet 2006).

If one refers to the traditional classification of things (Malaurie and Aynes 2015), biological elements are akin to movable things, as opposed to immovable things that cannot move. Then, they are consumable things to the extent that their use for a determined scientific research can lead to the loss of a part of this resource.[13] Finally, they are non-fungible things because there are no two identical biological samples

---

[10]Institut National du Cancer (INCA), *Conservation et utilisation des échantillons tumoraux en cancérologie*. INCA also published *Charte éthique des tumorothèques* in 2006.

[11]Council of Europe, recommendation CM/REC (2016) six of the Committee of Ministers to Member States on research using biological material of human origin, adopted on 16th May 2016.

[12]Art. L. 1211-2 CSP. See also Thouvenin (2005), p. 33.

[13]Hence the need for an optimal allocation of biological samples to research, the scientific relevance of which is examined by independent ethic committees, e.g. in France, the committees for the protection of persons.

placed in collections given the genetic information contained by each of them and of which contributes to their singularity.

Then, the detachment implies the integration of biological samples into legal trade. Since legal commerciality is understood as "*all the legal acts a thing can be the subject of*" (Loiseau 2000; Galloux 1989), it must be distinguished from merchantable commerciality. This ability to circulate materializes in the provisions of the public health Code which regulates the "cession", [14] "import" or "export" [15] of biological elements once they have been collected from a full body. Biological elements are things that circulate, which are subjects of contracts,[16] at a price fixed by the public person or not, and which are likely to incorporate the patrimony of the legal or natural persons who administer it.

Finally, the professional who has carried out sampling, or at least the organization which employs them,[17] has a hold on the material thing that constitutes biological samples. That right on the thing was conferred upon him by the prior consent process. Professionals are thus in a position of first possessor and first person who disposes of the collected element.

Biological elements, which have emerged at the end of a detachment process, are considered by law as things, but their transformation is not completed just yet.

2.2 The Transition from Biological Elements to Biological Samples

If biological elements are considered as things integrated in the legal trade, researchers base their experiments on biological samples, which are new goods resulting from the combination of biological elements and the know-how of pathologists. The emergence of biological samples is akin to a specification process (Sect. 2.2.1) which, although useful in determining the owner of the samples, does not abolish the unavailability inherent in those goods of human origin (Sect. 2.2.2).

2.2.1 A Specification Process Turning a Thing into a Good

According to a classic formula, if all goods are things, all things are not goods. While many definitions have been put forward to qualify a good, none of them are unanimous in doctrine (Grzegorczyk 1979). We shall focus on the distinction criteria

[14]Art. L. 1243-4 CSP.

[15]Art. R. 125-5 CSP.

[16]Within the meaning of Article 1101 of the Civil Code.

[17]Generally, a public or private health institution, a university institution or a pharmaceutical institution.

proposed by Professors Zenati and Revet, according to whom a good must be understood as an "identifiable and reportable entity, with usefulness, being the purpose of an exclusivity relationship" (Zenati-Castaing and Revet 2008). In the present case, the usefulness (Jean 2012) of biological elements refers to the benefits that it is likely to provide in a therapeutic and scientific context. This usefulness is closely linked to the value of biological elements: a scientific, therapeutic and even economic value. If the usefulness of a thing is also related to its rarity, it will only become apparent in the case of biological sampling after the technical work carried out by professionals.

In fact, the use of "raw" biological elements collected from a full body will only be possible after a thorough exploitation of its informational content. Through their work, pathologists will reveal the characteristics of each sample, extract derivatives, put them into culture, or even duplicate them. Through their choices and conditioning processes, they will develop a quality "finished product", which could be made available to the entire scientific community. It is through this work which is akin to a specification process, as described in Articles 570 and 571 of the Civil Code,[18] that the transformation of biological elements into biological samples will be carried out, from a thing to a good. The industry would represent the main part, the biological element being its support and accessory. The actual value added comes from the work of professionals and not from biological elements themselves.

After that step, biological samples have little in common with the initial biological element. On the one hand, their form is different since they will generally be reconditioned in various ways (cryopreservation, paraffin block, microarray tissue, culture for duplication . . .) to meet the needs of researchers. On the other hand, they also differ in substance since the work of pathologists will have allowed them to extract information and to know their contents better (DNA, RNA, percentage of tumor cells, type of tumor cell . . .).

Approached from the point of view of specification, biological sample would therefore be a new good, distinct from the initial biological element. Such an approach would make it possible to answer the thorny question of ownership of biological samples. They are the fruit of the work (Revet 1991) of specifiers on biological elements, who would be considered as the owners of the biological samples, or the organization that employs them, in return for a donor compensation whose form remains to be determined. Such compensation could not be economic in nature because of the principle of non-patrimony of the human body which prohibits any economic profit from the elements of someone's own body. However, as already

[18]Art. 570 of the Civil Code: "*If an artisan or any person used a material that did not belong to them to form a thing of a new kind, either that the material may or may not return to its initial form, the one who owned it has the right to claim the thing that has been formed by refunding the estimated labour force price on the date of reimbursement* ». Art. 571 C. Civ.: "*If, however, the labour force was so important that it went well over the value of the material used, the industry would then be deemed the main part, and the worker would have the right to retain the thing they had worked on, by reimbursing the owner of the price of the material estimated on the date of reimbursement*".

provided for in some *Material transfer agreements*, feedback to donors on research carried out on their samples, a priority of access to innovative therapeutics developed from their samples or a co-ownership on certain patents issued is possible (Bellivier and Noiville 2006).

2.2.2 A Conditional Use of Biological Samples

Since the nature of biological samples is now acquired, we must next determine the conditions for the exercise of ownership right over these atypical goods and the modalities of their valorization. In this study, we understand the term "valorization" as any action to increase the quantitative and qualitative value of biological samples.

The property as defined by the Civil Code is understood as "the right to enjoy and dispose of things in the most absolute way".[19] However, the classical attributes of *usus, abusus* and *fructus* are not unanimously recognized in doctrine (Dross 2014), and are only partially applicable to biological specimens because of their human origin. Thus, the owner has goods, the availability of which is limited in many ways.

A limitation of biological samples availability is first related to the protection of the person requirements. The principle of withdrawal of consent[20] diminishes the owner's right on their goods, knowing that the person at the origin of the donation will be able to request the cessation of research on their samples as well as their destruction.[21] This possibility could weaken investment projects in the field of biological samples. Analogously, like the image of a bank whose customers would all withdraw their savings at the same time, biobanks incur a risk of "bankruptcy" if all donors, in the absence of trust in the institution, ordered the destruction of stored samples.

The protection of persons lending itself to research also involves contacting the patient[22] in case of change of purpose[23] in the use of the samples or a substantial modification of the conditions of research.[24]

The limitations on the availability of biological samples are then linked to public policy imperatives. This is particularly the case of administrative procedures prior to any activity of preparation and preservation of elements and products of the human body for scientific use or cession.[25] The creation of a biological samples collection is

[19]Art. 544 Civil Code.

[20]This principle is reflected in Articles L. 1211-2 and L. 1241-1 of the CSP.

[21]Unless there is an irreversible anonymity of the samples.

[22]Obtaining a new express consent for genetic research and non-objection of the patient for other types of research.

[23]Transition from a therapeutic purpose to a scientific purpose or vice versa.

[24]Art. L. 1123-9, R1123-42 et seq. CSP.

[25]Art. L. 1243-3 and L. 1243-4 CSP.

also subject to the approval of an ethic committee[26] which evaluates the scientific relevance of the research projects on which collections are based.

There are also health and safety requirements that limit the owners' claims. Thus, the destruction of elements or products of the human body is subject to compliance with the rules on surgical waste.[27]

Finally, the owner's power of disposition will also be limited by the circulation rules applicable to biological samples. Cessions of biological samples shall be carried out in France only between organizations which have obtained administrative authorization to carry out this type of operation. The import and export of biological samples shall also be subject to an administrative authorization issued by the Ministry of Higher Education and Research, and, for certain types of samples, an authorization from the Biomedicine Agency is also needed.[28] Inter-state trade will have to comply with the safety rules for the international transport of biological materials. A special authorization from the ministry will have to be obtained in case of exchanges with third countries to the European Union.

It has become all too clear that biological samples, derived from a process of specification, belong to the category of goods. As such, they remain subject to appropriation and are placed within legal trade. However, this situation is not such as to hinder the cardinal principles of protection of persons who lend themselves to scientific research entrenched in law, as one can note the evidence of numerous limitations on the availability of biological samples.

References

Bayer, E. (2003). *Les choses humaines*. PhD, Toulouse.

Bellivier, F., & Noiville, C. (2006). *Contrats et vivant* (p. 268). Paris: LGDJ.

Binet, J. R. (16th February 2017). Respect et protection du corps humain – Eléments et produits du corps humain – Thérapies génique et cellulaire : médicaments de thérapie innovante et préparations. – Sang humain, *J.-Cl. Civil Code.*

Bioy, X. (2010). *Vers une politique publique des « biobanques », RDSS* (p. 885).

Carbonnier, J. (2000). *Droit civil, Les personnes* (p. 19). Paris: PUF.

Cornu, G. (1997). *Droit civil, Introduction, les personnes, les biens* (8th ed., p. 168). Paris: Montchrestien.

Dross, W. (2014). *Droit des biens* (2nd ed., pp. 19–21) Paris: LGDJ.

Galloux, J.-C. (1989). Réflexion sur la catégorie des choses hors du commerce : l'exemple des éléments et des produits du corps humain en droit français. *Les cahiers du Droit, 30*(4), 1018.

Grzegorczyk, C. (1979). Le concept de bien juridique : l'impossible définition. *APD, 24*, 259.

Jean, S. (2012). *La protection des droits subjectifs par la responsabilité civile*. PhD, Toulouse (p. 157).

Loiseau, G. (2000). Typologie des choses hors du commerce. *RTD Civ.*, 48.

Malaurie, P., & Aynes, L. (2015). *Les biens* (6th ed., p. 62). Paris: LGDJ.

[26]Art. L. 1123-7 CSP.

[27]Art. R1335-1 et seq. CSP.

[28]Art. R1335-1 et seq. CSP.

Noiville, C. (1997). *Pour une protection des collections scientifiques*, *D* (p. 245).
Prieur, S. (1998). *La disposition par l'individu de son corps*. Th., Dijon, Les Études Hospitalières (p. 71).
Revet, T. (1991). *La force de travail : étude juridique*. PhD, Montpellier.
Rial-Sebbag, E. (2008). Genèse d'un cadre réglementaire pour les collections d'échantillons biologiques utilisées en recherche. Exploration d'un modèle de gouvernance. *RGDM, 27*, 233–271.
Thouvenin, D. (2005). Les banques de tissus et d'organes: les mots pour les dire, les règles pour les organiser. *LPA, 35*, 33.
Zenati-Castaing, F., & Revet, T. (2006). *Manuel de droit des personnes* (p. 232). Paris: PUF.
Zenati-Castaing, F., & Revet, T. (2008). *Les biens* (3rd ed., p. 21). Paris: PUF.

Chapter 3
Personal Data Protection in Tumor Banks

Emilie Debaets

The biological samples collected, registered and stored in biobanks could not be used for medical research without their associated personal data which can result in their direct or indirect identification. Any research on biological samples such as tumors—whose specificity lies in the fact that they were only removed as part of medical care—necessarily implies the collection and use of personal data.[1] The issue of personal data is "crucial among other issues which arise in biobanks management" (Rial-Sebbag 2008) and, *a fortiori,* in tumor banks management.

Personal data associated with biological samples by tumor banks are twofold (Thouvenin 2005). The first set of data is existing information collected during sampling relating to demographic information (name, address, date and place of birth, etc.), biological information (gender, ethnic background, genetic data, etc.), medical information (diagnosis, tests, therapy, personal or family history, diseases or related events, etc.), and, possibly, environmental information (family situation, professional life, sexuality, lifestyle and behavior, etc.). The second set of data is new information about the sample itself (nature of the lesion,[2] extension of the lesion,[3] etc.) and the conservation conditions (type of sample, method of preparation,

[1]CCNE, notice No. 25 on the application of genetic testing in individual studies, family studies and population studies (problems of DNA banks, cell banks and data computerization), June 24th, 1991; OECD recommendation on Biobanking and databases of human genetics research, October 22nd, 2009.

[2]The pathologist will assign a lesional code (ADICAP code) based on the method of sampling (biopsy, necropsy, surgical specimen, puncture, etc.), on the type of technique and on the organs or areas in question.

[3]The pathologist analyzes the extension of the tumors from the TNM system that classifies cancer stage. The system is based on the size of the primitive tumor and/or its extension to the neighboring tissues and organs (T), on the importance of the spread to the regional lymph nodes (N) and on the presence or absence of metastases (M).

E. Debaets (✉)
University Toulouse 1 Capitole, Maurice Hauriou Institute, Toulouse, France
e-mail: emilie.debaets@ut-capitole.fr

X. Bioy (ed.), *Public Regulation of Tumor Banks*,
https://doi.org/10.1007/978-3-319-90563-1_3

associated biological resources,[4] etc.) of analysis. This second category of information also constitutes personal data, notably data concerning health, the definition of which has recently been clarified by EU Regulation No. 2016/679.[5]

The quantity and quality of this data are likely to vary according to tumor bank policies. Firstly, regarding the quantity of data, variation does not seem to be related to the software used to manage activity but is mainly due to the protocols in place. Software can then be subject to specific technical adjustments.[6] On the other hand, regarding the quality of this data, variation also arises from the implemented protocols. The degree of annotation accuracy at the time of the actual sampling or during its transmission can be different; updating data may also vary according to the interoperability established with the patient's record.[7] The fact that tumor banks belong to networks leads to the definition of a quantitative and qualitative minimum standard.[8] Nevertheless, they can always record more data and include more details for scientific and/or economic reasons.

The conditions in which all this data can be used are determined by a set of French and European statutes, which are binding or non-binding. There are no specific binding statutes on biobanking in French or European law.[9] The conditions under which personal data may be used are therefore determined by general statutes. At a national level, they are a combination of the provisions of the French Data Protection Act[10]—especially chapter IX, which relates to the processing of personal data for research purposes in the field of health—and certain provisions of the Public Health Code.[11] At the European level, they stem from the provisions of EU Regulation

[4]These associated biological resources can be serum, plasma, blood or other liquids such as urine.

[5]They are defined by Article 4, 15 °, of this Regulation as all the information "relating to the physical or mental health of an individual, including the provision for health care services, which reveal information about this person's health." For more information, see § 35.

[6]Among existing software, see for example Tumorotek, TDBiobank, Modulbio, on the diversity of choices made by tumor banks for their equipment (commercial software, academic software, a combination of both or other) see INCA, *Données des rapports d'activités des tumorothèques*, years 2010-2011, March 2014, 56 p., esp. p. 11.

[7]Ibid., esp. p. 11.

[8]Ibid., esp. p. 12.

[9]The OECD has adopted non-binding guidelines for biobanks: OECD, *Guidelines for Human Biobanks and Genetic Research Databases*, October 22nd 2009.

[10]Act No. 78-17 of January 6th 1978 includes specific provisions regarding the processing of personal data for research, study or assessment purposes in the field of health. These provisions, established by the Act of July 1st 1994, No. 94-548 concerning the treatment of personal data for research purposes in the field of health and amending Act No. 78-17, have been recently changed by section 193 of the Act No. 2016-41 of January 26th 2016 which modernized our health care system.

[11]See for example Art. L. 1110-4 and L. 1111-8 in the Public Health Code.

No. 2016/679—which will replace the current directive (No. 95/46/EC) in May 2018[12]—and from the provisions of ETS Convention No. 108.[13]

Unlike the statutes governing the use of biological samples,[14] the various statutes governing the use of personal data do not specifically address the questions of their legal status or the nature of the rights on personal data. However, these questions are increasingly addressed and debated (Eynard 2013, p. 155) with more discussion on the idea of patrimony—even for data concerning health (Cavalier 2016)—which the French Council of State has strongly opposed (French Council of State 2014, p. 264).

Whatever the underlying qualifications, these statutes allow the use of personal data within a protective framework for data subjects. Moreover, this framework is sometimes adapted to the sensitivity of the data—as is the case with health data—and to the specificity of purpose—as is the case with research, study and assessment purposes in the field of health. Thus, while personal health data may not, in principle, be used due to their high sensitivity,[15] there are exemptions including when their use is justified by public interest, when it is for health purposes or when it is for research, study and assessment purposes in the field of health.

Tumor banks, as "disease biobanks" (Bellivier and Noiville 2009, p. 33), i.e. as biobanks of specialized research devoted to the study of the processes specific to each carcinogenesis, use personal data, including personal health data, for research, study or assessment purposes in the field of health. This use might be permitted, but it is constrained. Tumor banks must meet the legal conditions laid down by French and European statutes regarding the rights of the data subjects and the obligations of those who process or determine the processing of personal data. The French Data Protection Act further includes the requirement for prior authorization by the CNIL or a declaration of conformity to one of the two reference methodologies recently adopted.[16] Beside the particular context of tumor and personal data collection for a well-defined research project, these methodologies do not apply to the activities of tumor banks, which remain subject to the requirement of prior authorization.

[12]EU Regulation 2016/679 of the European Parliament and of the Council of April 27th, 2016 on the protection of natural persons with respect to the processing of personal data and the free movement of such data, and repealing directive 95/46/EC; Directive 95/46/EC of the European Parliament and of the Council of October 24th, 1995, on the protection of natural persons with respect to the processing of personal data and the free movement of such data.

[13]Convention ETS No. 108 for the protection of persons against the automated treatment of personal data, January 28th, 1981.

[14]Art. 16-1 in the Civil Code.

[15]Art. 8 of Act No. 78-17; Art. 8 of Directive 95/46/CE; Art. 9 of Regulation (EU) 2016/679; Art. 6 of Convention ETS No. 108.

[16]CNIL, delib. No. 2016-262, July 21st, 2016, *Méthodologie de référence MR-001—Recherches dans le domaine de la santé avec recueil du consentement*; CNIL, delib. No. 2016-263, July 21st, 2016, *Méthodologie de référence MR-003—Recherches dans le domaine de la santé sans recueil du consentement*.

Nevertheless, these methodologies will apply to the processing of personal data implemented later as part of the programs or research projects.

The use of personal data by the tumor banks mainly faces two types of constraints—both ethical and legal (Bellivier and Noiville 2009, p. 55)—which are expressed in French and European statutes with more or less clarity despite the recent adoption of the European General Data Protection Regulation (GDPR) and the French law of modernization of our health care system. Is personal data collected with the consent of the data subjects, and is it used with due respect for their rights and freedoms? The analysis of the French and European statutes shows that consent is an uncertain requirement for the collection of personal data (Sect. 3.1), and confidentiality seems to be an illusory requirement when it comes to the circulation of such data (Sect. 3.2).

3.1 Consent as an Uncertain Requirement for the Collection of Personal Data by Tumor Banks

Unlike the French Bioethics Act, the French Data Protection Act was not originally based on the requirement for consent of the data subjects. However, this requirement has gradually permeated the French Act under the influence of European law (Lepage 2005 [2015], p. 231). Consent is even a central issue as evidenced by Article 8 § 2 of the Charter of Fundamental Rights of the European Union. However, many uncertainties surrounding the necessity and modalities of consent as well as its extent and effects remain to this day.

3.1.1 Uncertainties Surrounding the Necessity and Modalities of Consent

The necessity for tumor banks to obtain consent from data subjects for the collection of their personal data is approached by French and European statutes in different ways.

Article 9 of the GDPR provides two exemptions on the collection of personal data by tumor banks: one is specifically on grounds of public interest in the field of public health[17]; the other is more generally on grounds of scientific research.[18] These two exemptions involve the existence of appropriate and specific measures to safeguard the fundamental rights and interests of the data subjects. Therefore, the collection of

[17]Art. 9 i and § 53 of EU Regulation 2016/679.

[18]Art. 9 j of EU Regulation 2016/679.

personal data could happen without their consent.[19] Much like Directive No. 95/46 (Schulte In Den Baümen et al. 2010), the GDPR finally gives Member States leeway to determine appropriate and specific measures.

Chapter IX of the French Data Protection Act includes a specific regime for research, studies and evaluations in the field of health that is based on the consent of the data subjects. This consent has special modalities. Between the opt-in model (express consent) and the opt-out model (presumed consent), the French Act has opted for the latter. As long as the data subjects are duly informed and do not oppose it, for whatever reason, their personal data may be collected.[20] Similarly, unless deceased individuals had expressed their refusal in writing in their lifetime, it is possible to collect their data.[21] The benefit of such a model is obvious: "It allows [for the recovery of] data from those who have remained in passive agreement and, at the same time, to increase the pool of resources" (Bellivier and Noiville 2009, p. 61). By facilitating the collection of personal data, the French Act encourages and supports research in the field of health.

Nevertheless, the opt-out model is no less subject to criticism in that it would not be genuine consent. To use personal data, the French Data Protection Act "does not require the agency or the researcher to ask the person if they agree or not; it merely imposes to inform the person of their right to refuse for their data to be processed" (Thouvenin 2005). Such deformation of consent "can be justified for reasons related to scarcity or urgency, which is the case with organ transplant, but the same justifications do not seem as obvious when it comes to constituting a pool of data for research" (Bellivier and Noiville 2009, p. 62).

This model of presumed consent could be challenged by the GDPR, which defines consent as "a clear affirmative act establishing a freely given, specific, informed and unambiguous indication of the data subject's agreement…silence, pre-ticked boxes or inactivity should not therefore constitute consent".[22] However, as this has been highlighted above, the GDPR does not impose to collect the data subject's consent when personal data is used for public health or scientific research purposes. Thus, it is difficult to conclude with certainty, in the absence of a legal interpretation of these provisions, what sense should be given to their combination. The presumed consent adopted by the French Act could be interpreted as being contradictory to the definition of consent. However, another interpretation could see it as a special guarantee provided by national law, in an area where the GDPR does not require to obtain consent and therefore give leeway to national law.

---

[19] § 54 of EU Regulation 2016/679 regarding more specifically the grounds of public interest in the field of health.

[20] Art. 56 al. 1 of Act No. 78-17.

[21] Art. 56 al. 3 of Act No. 78-17.

[22] § 32 of EU Regulation 2016/679.

3.1.2 Uncertainties Surrounding the Extent and Effects of Consent

Tumor banks are also facing the difficulty of determining the extent of consent as far as we do not know what future research on tumors and associated personal data will be like (Bellivier and Noiville 2009, p. 63).

Beyond the reversibility of the arguments relating to the development of scientific research and the protection of data subjects (Coulibaly 2011, p. 537), open consent can develop due to the vagueness of the statutes. Recognizing that it is not always possible to fully identify the purpose of scientific research at the time of data collection, the GDPR seems to allow open consent for research in public health, such as cancer research, as long as ethical standards in scientific research are met.[23] The French Data Protection Act has waived the issue by retaining the principle of presumed consent. Thus, biological samples and personal data stored in tumor banks can be used by any interested researcher as long as the data subjects do not object and that the research project does not imply a review of genetic characteristics.[24]

However, this question of open consent or specific consent necessarily arises through the obligation to provide information. The GDPR gives leeway to researchers. It provides exemptions to the obligation to provide information when it is impossible to do so or when it requires a disproportionate effort which would render impossible or jeopardize the achievement of the objectives of the processing.[25] Similarly, the French Act proves to be receptive to researchers. On one hand, in instances of difficulty finding relevant data subjects, the CNIL can authorize research on them without them being informed. The French Act has not restricted the scope of this exemption by defining the criteria that may be considered (e.g. financial means, technological means, time, number of people concerned, age of data, etc.), but the CNIL makes it part of a strict implementation (Coulibaly 2011, p. 562). On the other hand, public sector research projects on data that is not directly identifiable have also been recently allowed to waive this obligation to provide information.[26]

The extent of consent is uncertain, and its effects are not determined, either. Certainly, French and European statutes provide for a right to deletion "that is, somehow, the logical corollary and the precursor of the right to withdrawal" (Bellivier and Noiville 2009, p. 66). However, this right to deletion of personal data poses a few problems (Bellivier 2009, p. 131).

[23] § 33 of EU Regulation 2016/679.

[24] Art. 56 al. 2 of Act No. 78-17.

[25] Art. 14 of EU Regulation 2016/679.

[26] On the implemented special scheme, see Art. 57 III of Act No. 78-17 and Art. 36 of Decree No. 2016 - 1872 to December 26th, 2016, amending Decree No. 2005-1309 of October 20th, 2005 taken for the application of Act No. 78-17.

Firstly, the right to deletion has long been regarded as conditioned to the inaccuracy or the ambivalence of personal data.[27] Indeed, the French Data Protection Act perpetuates uncertainty. The right to object is an absolute right which can be applied without justification.[28] Therefore, the exercise of this right should involve the deletion of data without condition. Chapter IX, however, provides no details on the issue. While uncertainties remain in the French Act, the GDPR now clarifies this right. It expressly provides that the data subject may request the deletion of their data when they withdraw their consent or when they choose to object.[29]

This right to deletion could nevertheless represent a barrier to research when persons with very rare characteristics object, because they then limit research potential. Thus, this right is disputed in medical research. Therefore, the GDPR provides the possibility of exemption in the fields of public health and scientific research. The exercise of this right, however, remains a largely academic hypothesis for tumor banks because of the status of the data subjects (Coulibaly 2011, p. 538). As they are cancer victims, it is rare for them to oppose the collection of their personal data for medical research.

Suffice it to say that the consent requirement of the data subjects to the collection of their personal data by tumor banks raises many questions left unresolved by French and European statutes. However, the uncertainty surrounding this requirement is not the only source of difficulty related to the activities of tumor banks. The requirement for data confidentiality during transmission to research projects is also an issue.

3.2 Confidentiality as an Illusory Requirement for the Transmission of Personal Data by Tumor Banks

As personal data collected and associated with biological samples by tumor banks are essential for cancer research, they must be able to circulate between researchers. Considering the need to allow the development of medical research, French and European statutes allow the transmission of personal data. But this transmission, is regulated. It implies, with some exceptions, a requirement of confidentiality of personal data. However, confidentiality is an illusory requirement for the transmission of personal data because it is only relative (Sect. 3.2.1) and eventually turns out to be rather artificial (Sect. 3.2.2).

[27] Art. 40 of Act No. 78-17; Art. 12 of Directive 95/46/EC.

[28] Art. 56 al. 1 of Act No. 78-17.

[29] Art. 17 of EU Regulation 2016/679.

3.2.1 A Relative Requirement for the Transmission of Personal Data

French and European statutes allow tumor banks to transmit the personal data they hold to research teams as long as they guarantee their confidentiality.

The GDPR certainly does not expressly refer to a confidentiality requirement. It does require, on one hand, that the processing of personal data on grounds of public interest in the field of public health should be subject to "specific and appropriate measures" such as professional secrecy.[30] On the other hand, it requires that the processing of personal data for scientific research purposes have "appropriate safeguards" through the implementation of "technical and organizational measures" such as pseudonymization.[31] Thus, confidentiality is an underlying requirement. In addition, the GDPR also puts an end to the possibility for researchers to waive the established protection by arguing that the data used does not include the names, addresses, social security numbers, etc. of the data subjects, as it was notably the case in the United Kingdom and Greece (Rumbold and Pierscionek 2017). In the continuity of the analyses of the G29,[32] the GDPR has specifically clarified the status of this pseudonymized data[33] to ensure better protection of data subjects and greater confidence in the research.

The French Data Protection Act is expressly based on this confidentiality requirement. The conditions under which personal data may be transmitted have recently been simplified by the law of modernization in our health care system. They no longer refer to the obligation to code data, nor to exceptions and their uncertain scope (Coulibaly 2011, p. 421). The law now provides that the transmission of personal data is possible only if there are guarantees ensuring its confidentiality.[34] Will these guarantees be assessed according to the nature of the research project for scientific and/or technical reasons (that are also the expression of economic interests) (Ponthoreau 1996)? To avoid any abuse, these guarantees should therefore make no exceptions.

The requirement of confidentiality provided for by French and European statutes does not encompass the anonymity of personal data, i.e. an absolute impossibility for the identification of the data subjects. Such anonymity could be an impediment to research projects by preventing researchers from distinguishing one case from another, making correlations, and/or checking research results. The confidentiality requirement is only a pseudonymization of data, i.e. a relative impossibility to identify the data subjects. Pseudonymization may involve different techniques: a

---

[30] Art. 9 i. of Regulation (EU) 2016/679.

[31] Art. 89 of Regulation (EU) 2016/679.

[32] G29, Opinion 04/2007 on the concept of personal data, esp. pp. 19–21; Opinion 05/2014 on Anonymisation Technique, esp. p. 11.

[33] Art. 4 § 5 of EU Regulation 2016/679.

[34] Art. 55 of Act No. 78-17.

cryptographic system with a secret key, a hash function (with salt, by key with a recorded key, by key with key deletion), or tokenization.[35] Here, once again, the balance between the interests of researchers and those of the data subjects will be based on the doctrine of the CNIL (Ponthoreau 1996), the French Data Protection Act having merely provided this possibility of transmission and delegated the implementation with the possibility to adopt recommendations or references in this area.[36]

3.2.2 An Artificial Requirement for the Transmission of Personal Data

Despite the technical means used to ensure confidentiality, the risk of re-identification through new processes of aggregation and data cross-checking cannot be excluded.

These technical means do not reduce the risks for the data subjects, but simply reduce the probability of their occurrence. If data can still be linked to a specific natural person, the risks to the data subject will be of the same nature and the same scope: processing may give rise to discrimination and data subjects might be deprived of their rights and freedoms, especially of their right to privacy. Several recent studies have clearly highlighted these risks of re-identification (French Council of State 2014, p. 312). For example, a French study has shown that 89% of patients who had a hospital stay during a given year could be identified only by the following information: the hospital where they stayed, their postal code, their month and year of birth, their gender, their month of release and the duration of their stay. According to another study, 80% of Americans can be identified by the combination of their zip code, gender and date of birth. Therefore, a risk of re-identification can never be absolutely excluded in the context of biobanks despite the requirement of confidentiality (Milius et al. 2014).

In recent years, re-identification processes have made significant progress, causing the technical instruments previously regarded as reliable to be questioned. Therefore, should not the risk of re-identification through new processes of aggregation and data cross-checking be considered, even assuming we use rigorous technical instruments? The social acceptability of tumor banks would therefore not deny the persistence of these risks for the data subjects, as evidenced by the development of self-regulatory measures established by some of these structures, e.g. the International Cancer Genome Consortium (Milius et al. 2014; Joly et al. 2016).

[35] For a description of techniques, see G29, Opinion 05/2014 on Anonymisation Technique, esp. p. 22.

[36] Art. 55 par. 2 of Act No. 78-17.

For confidentiality not to be an artificial requirement, it should necessarily be accompanied by several measures. Firstly, an impact study should be carried out on current technical instruments and periodic reviews of this study should be conducted. As highlighted by the G29, the re-identification of data subjects is a growing threat and there is a grey area where it is difficult to measure if re-identification might be possible in advance.[37] This impact study and its periodic revisions could be based on three criteria—individualization, correlation and inference (as defined by the G29[38])—which have been adopted by the CNIL.[39] Anyway, an impact study is part of the new requirements of the GDPR.[40] Secondly, people should be informed of the risk of re-identification during data collection. The communication of personal data breaches will now be mandatory,[41] but this information, communicated a posteriori, remains insufficient. It should be communicated a priori, which is not specifically provided despite the long list of required information to provide.[42] These examples of additional measures highlight, when they are necessary, the possibility and the interest for a realistic approach for the confidentiality requirement.

To conclude, these two ethical and legal constraints faced by tumor banks—consent to collection of personal data and confidentiality during transmission—show the difficult and unavoidable conciliation between the need to allow the development of research on one hand and the need to ensure the protection of the people and maintain their trust in biobanks, such as tumor banks, on the other hand. In the absence of a conciliation that is predetermined by French and European statutes, it is then up to supervisory authorities, in particular to the CNIL and the G29, and the tumor banks themselves, to participate in the definition of this conciliation.

References

Bellivier, F. (2009). Le droit de retrait en bioéthique sur la voie de l'émancipation. *Droits, 48*, 131.

Bellivier, F. & Noiville, Ch. (2009). Les biobanques, PUF, collection Que sais-je ? 128 p.

Cavalier, M. (2016). *La propriété des données de santé*. PhD thesis, Université Lyon III.

Coulibaly, I. (2011). *La protection des données à caractère personnel dans la recherche scientifique*. PhD thesis Université de Grenoble, 1117 p.

Eynard, J. (2013). *Les données personnelles. Quelle définition pour un régime de protection efficace?* Michalon, 435 p.

[37] G29, Opinion 03/2013 on purpose limitation, esp. p. 31.

[38] G29, Opinion 05/2014 on Anonymisation Technique, esp. p. 13.

[39] See for example CNIL, Decision no. 2017-013 of 19th January 2017 authorizing Public Assistance—Hospitals of Paris to implement an automated treatment of personal data the purpose of which is a healthcare data warehouse, referred to as "EDS."

[40] Art. 35 of Regulation (EU) 2016/679.

[41] Art. 34 of Regulation (EU) 2016/679.

[42] Art. 13 of Regulation (EU) 2016/679.

French Council of State. (2014). *Le numérique et les droits fondamentaux*. La documentation française, 446 p.

Joly, Y., de Vries-Seguin, E., Chalmers, D., Ouellette, B. F., Yamada, J., Bobrow, M., et al. (2016). Analysis of five years of controlled access and data sharing compliance at the international cancer genome consortium. *Nature Genetics, 48*(3), 224.

Lepage, A. (2015). Consentement et protection des données à caractère personnel. In J. L. Girot (Ed.), *Le harcèlement numérique* (Vol. 2005, p. 231). Dalloz.

Milius, D., Dove, E. S., Chalmers, D., Dyke, S. O., Kato, K., Nicolas, P., et al. (2014). The international cancer genome Consortium's evolving data-protection policies. *Nature Biotechnology, 32*(6), 519.

Ponthoreau, M. C. (1996). La protection des personnes contre les abus de l'informatique. A propos de la loi du 1er juillet 1994 relative au traitement des données nominatives ayant pour fin la recherche dans le domaine de la santé. *RFDA*, p. 796.

Rial-Sebbag, E. (2008). Genèse d'un cadre règlementaire pour les collections d'échantillons biologiques humains utilisées en recherche. Exploration d'un modèle de gouvernance. *Revue Générale de Droit Médical, 27*, 233.

Rumbold, J. M. M., & Pierscionek, B. (2017). The effect of the general data protection regulation on medical research. *Journal of Medical Internet Research, 19*(2), e47 from p. 1.

Schulte In Den Baümen, T., Paci, D., & Ibarreta, D. (2010). Data protection and sample management in biobanking – A legal dichotomy. *Genomics Society and Policy, 6*(1), 33.

Thouvenin, D. (2005). Les banques de tissus et d'organes: les mots pour le dire, les règles pour les organiser. *Les Petites Affiches, 35*, 3.

Chapter 4
The Policies of Tumor Biobankers: Main Strategies and an Example of the Policies Adopted by the Nice Hospital Biobank, France

Paul Hofman

4.1 Introduction: Present and Future Issues

Biobanks (centers), which include tumor banks, are essential sources of biological materials for medical research. However, the expansion of fields such as translational research and especially personalized medicine and immunotherapy has heightened the need for biobanks to be run effectively and efficiently (Washetine et al. 2017a). In fact, ever-more research projects, both academic and commercial, are employing bioresources, thereby greatly increasing the demand for samples and associated clinical data from cancer patients. These bioresources are vital to research aimed at defining diagnostic biomarkers that are prognostic or predictive of therapeutic response, or at understanding the mechanisms involved in carcinogenesis.

Increased media attention and easier access to information has raised public awareness of the role biobanks play in modern medical research. Consequently, patients, patients' families, and patient associations have become stakeholders in the organization of biobanks and can play a role in discussions between physicians and biobanks about the creation and use of collections. Biobanks now have a wide range of stakeholders, many of whom have contributed to discussions on how collections of bioresources obtained from cancer patients should be managed (Mabile et al. 2013). The demand for collections of human samples is such that more than 2000 biobanks have been set up around the world. Europe's biobank infrastructure, the BBMRI (www.bbmri-eric.eu), alone lists at least 400 biobanks (Yuille et al. 2008). Consequently, commercial and academic research teams have a choice of sources for the biological samples they need. Key considerations for research teams when choosing which biobank(s) to work with are sample quality, which is strongly linked

P. Hofman (✉)
Laboratory of Clinical and Experimental Pathology, Nice Hospital Biobank (BB-0033-00025), University Hospital Federation OncoAge, Pasteur Hospital, Nice, France
e-mail: Hofman.p@chu-nice.fr

X. Bioy (ed.), *Public Regulation of Tumor Banks*,
https://doi.org/10.1007/978-3-319-90563-1_4

to handling and storage practices during the pre-analytical phases (Doucet et al. 2016; Freedman et al. 2015), and the availability of associated clinical data that are appropriate to the project. In fact, the quality of both samples and associated clinical data can affect a study's reproducibility, and this can have a major impact on research budgets, notably in the case of commercial research (Doucet et al. 2016; Freedman et al. 2015). Nevertheless, biobanks are expensive to run and their budgets can be difficult to balance, so a biobank's sustainability depends largely on its ability to cover the costs engendered by its work (Yuille et al. 2017). Although there are various keys to success for a biobank, as will be discussed below, the most important factors are implementing effective policies and adopting a well-thought-out strategy that incorporates the flexibility needed to quickly adapt to researchers' needs.

In general, it is imperative for a biobank to introduce policies which allow it to: (1) develop public-public and/or public-private partnerships, (2) maintain a high profile with local, national, and international institutions, notably by joining consortiums and research networks of excellence, and (3) follow a robust economic model and attain a stable budget. In this regard, it is important to reduce working constraints, while respecting patients' rights, ethics, and the relevant national legislation (Hofman et al. 2014).

Biobanks are evolving, as is shown by the emergence of what are known as "Next Generation Tumor Biobanks" (NGTB). In addition, some research activities are being dissociated from healthcare (diagnostic, prognostic, targeted therapy, immunotherapy) and from the development of genomic medicine (Clément et al. 2017). Many biobanks are taking on new roles, becoming centers of expertise in a wide variety of fields (clinical, pathology, genomics, bio-statistics, bio-informatics, and even mathematics), as well as repositories for the collection and storage of samples. Biobanks provide researchers with samples and associated clinical data enriched by specialists in a variety of fields, whose expertise gives added value to patient care, particularly through the discovery of new biomarkers and the validation of molecular targets for novel therapies. Hence, biobankers are not just bioresource warehouse managers but full partners in developing healthcare and research.

4.2 What Policy(ies) for a "Biobanker"?

The first priority for a newly created biobank is to define a policy and a strategy, which can then be adapted in response to changes in research activity and researchers' needs. This policy must be approved by the strategic committee and presented by the biobank's operating director to the biobank's staff and its governing institution as part of the biobank's certification and/or accreditation processes. The policy is set out during management reviews, which are conducted at regular intervals to maintain the biobank's accreditation and/or certification. A biobank may have one or more policies, depending on the strategy chosen by its director, physicians, pathologists, biologists, and managers.

In general, policies may be drawn up according to:

4.2.1 The Collection's Target Pathology/Pathologies, If Any

Some tumor biobanks, especially those run by pathologists, are "generalist" biobanks in that they collect all solid cancers and hematological malignancies that are examined and used for diagnosis in pathology laboratories. The main advantage of this type of biobank is its ability to respond to requests covering a wide range of pathologies. However, the generalist approach also disadvantages, as it can lead to the scattering of effort and funding, hinder access to satisfactory clinical data (e.g., postoperative survival rates for patients), and result in insufficient staffing. It can also be difficult for a generalist biobank to provide expertise in all domains and to store more than a limited number of samples for each pathology.

"Specialist" biobanks, of which there are fewer, collect and store samples relating to the pathologies of a specific organ (or a maximum of two or three organs). This strategy allows the biobank's staff (pathologists, data managers, technicians etc.) to focus their time and effort on a small number of collections. By concentrating on a limited of number of pathologies, specialist biobanks can offer their partners and clients greater expertise. However, such biobanks often have only a small number of samples, especially when starting their collections, so they are unable to meet some potential clients' needs. In addition, they have fewer opportunities to collaborate with other organizations because they work on only one or two types of cancer.

4.2.2 The Type of Bioresource Collected

The development of new techniques and the discovery of new biomarkers (genomic alterations, tissue antigens, epigenetic modifications, cytokines, etc.) means it is now essential to collect samples of a variety of biological resources (tissues, cells, fluids) from the same cancer patient. Hence, from initially focusing on tissue samples, tumor biobanks have enlarged their collections to include other biological products (plasma, serum, whole blood, urine, cerebrospinal fluid, pleural fluid, primary cells, feces, etc.). The ability to preserve tissues in paraffin after fixation in formalin has allowed biobanks to reply to new types of request by research scientists. It has also become possible to prepare tissue microarrays (TMAs) from several hundred tissues fixed in paraffin, thereby opening new perspectives for establishing partnerships with academic and commercial research teams.

4.2.3 The Biobanker's Field of Expertise

In most cases, a biobank's strategy for managing and using its collection(s) is set by the biobank's director/head in conjunction with its governing institution, which must validate this strategy. Most strategies fall into one of two main categories. (1) Many biobanks are run as simple biorepositories, that is, as locations for receiving, storing, safeguarding, and supplying biological resources for research. This is often the case for generalist biobanks. Because biorepository-type biobanks carry out little research, their collections can be managed by a technician with no specific expertise in biomedical research. (2) Other biobanks, especially those run by a research scientist, pathologist, or biologist, contribute more actively to research projects and work in close collaboration with their academic and commercial partners (Ilie et al. 2015, 2017; Rakha et al. 2015). The biobank's expertise adds value to its collections and can be used to help researchers develop new projects. Of course, this expertise usually covers just one or two organs or pathologies, although the biobank may also store and safeguard samples relating to other pathologies. Some biobanks carry out their own research and set up technical centers that enable them to transform the samples they store into by-products such as nucleic acids and proteins.

4.2.4 The Hospital's or Institution's Policies and General Organization

Hospitals tend to be divided into large departments, generally referred to as poles in France. Most large French hospitals house several poles (e.g., biology-pathology, oncology, radiology-imagery), as well as research and innovation departments or centers. Because biobanks have a transversal mission, they are often set up as units within a biology-pathology pole or a research and innovation center. This mode of functioning generally depends on the biobank's economic model, which is based on a pricing grid that is designed to ensure it can cover the cost of creating, storing, and administrating its collections (Clément et al. 2014). This is a sensitive issue, particularly in the eyes of the public, who must be reminded that it is not the biological samples themselves which are given a market value but the work performed by the biobank team and clinicians. This work includes obtaining consent from patients/donors, registering, storing, and safeguarding samples, quality control, compiling clinical data associated with the sample, and supplying samples to clients (Clément et al. 2014). In addition, a biobank must assess the value of any expertise (e.g., with respect to a cancer pathology) and/or biological or clinical data it supplies along with a sample, so it can consider this added value when calculating an appropriate cost for supplying the sample (Clément et al. 2014). Approval must be obtained for all sample transfers and a contract stipulating each partner's responsibilities and duties must be drawn up. Such contracts often take the form of a Materiel Transfer Agreement (MTA). However, depending on administrative constraints (most notably the strictness of intellectual property requirements), discussions with partners can be complex and hamper certain decisions, particularly when dealing with commercial companies (Hofman et al. 2014). Consequently, the speed at which samples can be supplied to researchers depends, at least in part, on the parent institution's administrative and legal departments. Finally, a biobank's budget is contingent on the parent institution's policy, the amount of support it is prepared to give, and the involvement of the biobank's staff. Ideally, a biobanker will obtain a sufficiently large return on investment to ensure the biobank's long-term future and development (perpetuate employment, optimize equipment, improve the computing system, etc.).

4.3 Choosing a Policy for a Biobank

Tumor biobanks are gradually developing policies of excellence in response to competition within the sector, which has increased greatly due to the recent massive growth in the number of biobanks, both public and private. Hence, it is important to publish productivity and efficacy indicators (Hofman et al. 2013). The most important indicators concern the quality of the biobank's procedures (certification and/or accreditation) (Long-Mira et al. 2016; Washetine et al. 2017b), the cost of the samples and services it provides, and details of scientific collaborations. The amount a biobank costs its parent institution gives value to its collections and to the samples it transfers to researchers (Clément et al. 2014). What is more, publishing performance indicators should help remove the barriers that exist between actors working in the public and private sectors (Hofman et al. 2014). Although it is very important for the public and private sectors to be able to work together openly (Hofman et al. 2014), it is currently inconceivable for public biobanks to transfer biological samples to private organizations, which could then supply them to commercial clients without informing the biobanker of the samples' final destination or the objectives of the research in which they will be used (Cooreman et al. 2017).

4.4 The Policy Drawn Up for the Nice University Hospital Biobank (BB-0033-00025, Nice, France)

When it was created in 2006, Nice University Hospital's biobank, with support from the university's governing body and a strategic committee, immediately defined an operating strategy (Hofman 2006). This strategy covered several key areas:

1. Scope of the sample and clinical data collection.

   It was decided to collect samples and data for only a limited number of pathologies and organs (chest pathologies, melanomas, thyroid pathologies). This strategy enabled the newly formed biobank to concentrate on gathering a limited range of samples (frozen tumor tissue, formaldehyde fixed tissues, blood samples, plasma, germ-line and somatic DNA, urine), mainly from patients with lung cancer.

(a) The decision to specialize was taken in the light of the expertise of Nice University Hospital's clinicians, surgeons, and pathologists with respect to lung, skin, and thyroid pathologies.
(b) Focusing on a limited number of pathologies and organs enabled the biobank to quickly build up a collection of samples from patients at the hospital with full clinical records. Most notably, it was able to collate survival rate data in conjunction with details of the course of the disease and treatment regimes.

2. Technical center.

   A technical center was set up to provide additional services.

   (a) The biobank's technical center expanded progressively as it obtained equipment to (1) extract nucleic acids and check their quality, automatically prepare TMAs, and carry out molecular analyses (including new generation sequencing techniques), (2) carry out in-situ analyses of the expression of proteins or nucleic acids, and (3) digitize images of tissue lesions giving morphological information for each specimen.

3. Pricing policy.

   The biobank's pricing policy considers the rules, legal conditions, and strict ethics (contracts, authorizations to export samples) the biobank must respect.

   (a) From the beginning, strict procedures were followed to obtain consent for donating biological samples from all patients. These procedures are adapted as necessary to cover all the different situations in which samples might be taken (Hofman et al. 2010).
   (b) The pricing policy was based on assessments initially made by the French National Cancer Institute (www.e-cancer.fr) and then by a European consortium (Clément et al. 2014). Prices considered the cost of collecting and safeguarding samples, as well as any additional input from the biobank's experts and/or any sample processing/preparation work carried out by the biobank's technical center.

4. Quality control.

   The biobank immediately introduced a strict quality control system so it could rapidly obtain certification and accreditation to international standards.

   (a) The quality control system set up when the biobank was created allowed it to obtain S96-900 certification by 2010. In addition, the pathology and molecular biology laboratory associated with the biobank obtained ISO 15189 accreditation in 2013 (Long et al. 2013, Washetine et al. 2013). Certification and accreditation have been maintained since then.

5. Scientific collaboration.

   The biobank took steps to establish partnerships with academic and private research units, using indicators (e.g., publications, patents) to evaluate the productivity of these collaborations. Particular emphasis was given to integrating the biobank's projects into national and international networks of excellence.

(a) Targeting a limiting number of collections relating to cancers of just three organs (lungs, skin, thyroid) and the biobank team's expertise in these domains enabled the biobank to set up scientific collaborations with other research teams, especially in the private sector. This has resulted in the publication of numerous papers and the award of several patents, mostly in association with private companies (Ilie et al. 2015, 2017; Lassalle et al. 2011; Rakha et al. 2015).
(b) The biobank has also joined several specialist networks, notably because of projects supported by the French National Cancer Institute [malignant mesothelioma network, malignant melanoma network (www.inca.fr)]. Since 2015, the biobank has been part of an AVIESAN-approved network of excellence called the OncoAge University Hospital Federation, whose aim is to improve care, training, and research with respect to cancers in older people (www.oncoage.org). The biobank is also a member of the Institute for Research on Cancer and Aging, Nice (IRCAN, www.ircan.org).

6. Innovation.

   The biobank uses its facilities and expertise to develop innovative products and processes.

   (a) Finance has been provided to enable the biobank to carry out several projects to develop innovative solutions, including a radio-frequency (RFID chips) tracking system (project carried out in collaboration with IBM and ST Microelectronics, http://rfidjournal.com), and a method for encapsulating DNA samples at room temperature (project carried out in collaboration with the French company Imagen).

7. Training.

   The biobank has created a postgraduate course in biobanking to train future biobank managers.

   (a) The biobanker's role has evolved and become very specific. Since 2016, Nice's Côte d'Azur University has offered a master's degree program in biobanking, which includes modules on managing complex and massive datasets (http://www.univ-cotedazur.fr/masterbiobank). The program differs from the Catholic University of Lyon's biobanking master's course (Gormally et al. 2017) in that it focuses more on biostatistics and bioinformatics, and in the international nature of its teaching and internships.

4.5 Conclusion: One or Several Policies for Biobankers?

Because the approaches adopted by the heads of biobanks vary between centers and between institutions, different biobanks may follow very different strategies. Nevertheless, the use of biological samples is governed by mandatory rules and regulations and by universal ethical considerations. The strategy adopted by the head of a biobank must be dynamic, so the biobank can overcome the challenges of the future (Clément et al. 2017; Washetine et al. 2017a, b). In particular, it has become essential to master the increasingly complex and massive datasets associated with biological samples. Another of a biobanker's central concerns is to balance the budget so the biobank can continue operating within an increasingly competitive market (Henderson et al. 2017; Macheiner et al. 2017; Yuille et al. 2017). A well-controlled pricing policy is a key factor in achieving this objective.

Very high throughput analyses have the power to generate many thousands of items from a single sample. This, together with the explosion in the number of associated stratified or precision medicine projects, means that biobanks have had to rapidly introduce strict pre-analytical and accreditation norms, and consult with multidisciplinary experts to implement new procedures (Doucet et al. 2016). Given this context, biobankers must optimize their policies for managing the complex and massive datasets associated with biological samples. In fact, biobanks which try to collect very large numbers of samples can find it difficult both to carry out all the necessary analyses fast enough and to store the data produced by these analyses. Consequently, future generations of biobankers will need to master high throughput analysis methods.

References

Clément, B., Hofman, P., Desille, M., & Dagher, G. (2017). Réalités industrielles. La normalisation des techniques d'ingénierie en recherche biomédicale: l'exemple des centres de ressources biologiques. *Annales des Mines, 2*, 24–28.

Clément, B., Yuille, M., Zaltoukal, K., Wichmann, H. E., Anton, G., Parodi, B., et al. (2014). EU-US expert group on cost recovery in biobanks. Public biobanks: Calculation and recovery of costs. *Science Translational Medicine, 6*(261), 261fs45.

Cooreman, A., Bravo, E., van Gool, A. J., Puchois, P., Roehrl, H. A., & Hofman, P. (2017). Traceability and transparency should be mandatory for all human biospecimens. *Applied Clinical Trials, 2*, 1–7.

Doucet, M., Becker, K. F., Björkman, J., Bonnet, J., Clément, B., Daidone, M. G., et al. (2016). Quality matters: 2016 annual conference of the National Infrastructures for biobanking. *Biopreservation and Biobanking, 15*, 270–276.

Freedman, L. P., Cockburn, I. M., & Simcoe, T. S. (2015). The economics of reproducibility in preclinical research. *PLoS Biology, 13*(6), e1002165.

Gormally, E., Hardy, I., Caboux, E., Di Donato, J. H., Hainaut, P., & Hofman, P. (2017). Training the next generation of biobankers: A two-years Master's course in the management of biobanks. *Biopreservation and Biobanking, 15*(5), 438–450.

Henderson, M. K., Goldring, K., & Simeon-Dubach, D. (2017). Achieving and maintaining sustainability in biobanking through business planning, marketing, and access. *Biopreservation and Biobanking, 15*(1), 1–2.

Hofman, V., Ilie, M., Long, E., Washetine, K., Chabannon, C., Figarella-Branger, D., et al. (2013). Measuring the contribution of tumor biobanks to research in oncology: Surrogate indicators and bibliographic output. *Biopreservation and Biobanking, 11*(4), 235–244.

Hofman, P. (2006). The CHU-CRLCC-UNSA tumour/tissue bank of the nice area. *Medical Science (Paris), 22*(1), 21–25.

Hofman, V., Bonnetaud, C., Gaziello, M. C., Ilie, M., Lassalle, S., Butori, C., et al. (2010). The nice CHU biobank experience to collect patients' informed consent for research context (2004-2009). *Annales de Pathologie, 30*(5), 337–343.

Hofman, P., Bréchot, C., Zatloukal, K., Dagher, G., & Clément, B. (2014). Public-private relationships in biobanking: A still underestimated key component of open innovation. *Virchows Archiv, 464*(1), 3–9.

http://www.rfidjournal.com/article/view8303. French biobank secure specimens.

Ilie, M., Szafer-Glusman, E., Hofman, V., Long-Mira, E., Suttmann, R., Darbonne, W., et al. (2017). Expression of MET in circulating tumor cells correlates with expression in tumor tissue from advanced-stage lung cancer patients. *Oncotarget, 8*, 26112–26121.

Ilie, M., Nunes, M., Blot, L., Hofman, V., Long-Mira, E., Butori, C., et al. (2015). Setting up a wide panel of patient-derived tumor xenografts of non-small cell lung cancer by improving the preanalytical steps. *Cancer Medicine, 4*(2), 201–211.

Long, E., Hofman, V., Ilie, M., Washetine, K., Lespinet, V., Bonnetaud, C., et al. (2013). Accreditation of the activity of molecular pathology according to ISO 15189: Key steps to follow and the main potential pitfalls. *Annales de Pathologie, 33*(1), 12–23.

Long-Mira, E., Washetine, K., & Hofman, P. (2016). Sense and nonsense in the process of accreditation of a pathology laboratory. *Virchows Archiv, 468*(1), 43–49.

Lassalle, S., Hofman, V., Ilie, M., Butori, C., Bonnetaud, C., Gaziello, M. C., et al. (2011). Setting up a prospective thyroid biobank for translational research: Practical approach of a single institution (2004-2009, Pasteur hospital, nice, France). *Biopreservation and Biobanking, 9*(1), 9–19.

Mabile, L., Dalgleish, R., Thorisson, G. A., Deschênes, M., Hewitt, R., Carpenter, J., et al. (2013). BRIF working group: Quantifying the use of bioresources for promoting their sharing in scientific research. *Gigascience., 2*(1), 7.

Macheiner, T., Huppertz, B., Bayer, M., & Sargsyan, K. (2017). Challenges and driving forces for business plans in biobanking. *Biopreservation and Biobanking, 15*(2), 121–125.

Rakha, E., Pajares, M. J., Ilie, M., Pio, R., Echeveste, J., Hughes, E., et al. (2015). Stratification of resectable lung adenocarcinoma by molecular and pathological risk estimators. *European Journal of Cancer, 51*(14), 1897–1903.

Washetine, K., Ilié, I., Hofman, P., Butori, C., Long, E., Lassalle, S., et al. (2017a). Les biobanques : quels enjeux en 2017 ? *Revue Francophone Des Laboratoires, 448*, 25–28.

Washetine, K., Hofman, V., Lassalle, S., Long, E., Ilié, M., Butori, C., et al. (2017b). L'accréditation des laboratoires d'ACP : pourquoi est-ce incontournable? *Revue Francophone Des Laboratoires, 448*, 31–37.

Yuille, M., van Ommen, G. J., Bréchot, C., Cambon-Thomsen, A., Dagher, G., Landegren, U., et al. (2008). Biobanking for Europe. *Briefings in Bioinformatics, 9*(1), 14–24.

Yuille, M. M., Feller, P. I., Georghiou, L., Laredo, P., & Welch, E. W. (2017). Financial sustainability of biobanks: From theory to practice. *Biopreservation and Biobanking, 15*(2), 85–92.

Chapter 5
The Concept of Biological Samples Collection

Anne Brouchet and Sophie Peries

5.1 Introduction

Collections of biological tumor samples gather valuable and substantial information, essential for the development of biomedical research and crucial for a better understanding on cancer biology.

The establishment of a biological collection involving the human person and the compilation of the related clinical data are governed by legal and ethical rules regulating clinical research.

Collections management presumes strict organization, rigorous structuration, IT tools, funding and dedicated qualified personnel within a Biological Resource Centre (BRC). The main mission of a BRC is to ensure the reception, preservation, preparation and disposal of biological resources for the needs and satisfaction of all parties involved (researchers, industrialists) to guarantee the respect of the new regulation in force since November 2016 (Jardé Law) and ensure controlled biosecurity conditions (certification according to the French standard NF S96-900).

5.2 Definition of Biological Sample and Biological Collection

5.2.1 Biological Sample

"Human biological samples are defined as 'tissues and cells from human body and their derivatives...' (Article L. 1243-3 of the French Public Health Code)."

A. Brouchet (✉) · S. Peries
Biological Resource Centre (BRC) Cancer, Toulouse University Hospital, Toulouse, France
e-mail: brouchet.anne@chu-toulouse.fr; peries.s@chu-toulouse.fr

X. Bioy (ed.), *Public Regulation of Tumor Banks*,
https://doi.org/10.1007/978-3-319-90563-1_5

Several types of samples can be distinguished: DNA, RNA, serum, urine, feces, saliva, tissues, biopsies, cells, perspiration, semen, hair, nails...

These are obtained by collecting or sampling, and their origins are diverse (treatment, donation, swabs collected for the need of a biomedical research...).

5.2.2 Biological Collection

The Law no. 2004-800 of August 6th, 2004 on Bioethics provides a definition of a biological collection: "Gathering of biological samples for scientific purposes (tissues and cells from a human body and their derivatives, organs, blood and its derivatives) swabbed on a group of identified and selected persons according to clinical and biological characteristics from one or several members of the group, as well as the derivatives of these samples."

A collection always serves a purely scientific purpose (research) and not a medical one (treatment).

5.3 How Are Collections Established and Used in 2017?

5.3.1 Establishment of a Biological Collection

The establishment of new human biological collections as part of a research project must comply with the new regulation governing clinical research (Jardé Law; Application Decree no. 2016-1537). Three newly defined research categories are then applicable: interventional studies, interventional studies presenting minimal risks and constraints and non-interventional studies (see below).

In any case, research involving the human person including sampling and the collection of new biological specimens (blood, biopsies, swabs) require approval from an Ethics Committee, the Committee for the Protection of Persons (Comité de Protection des Personnes or CPP), in charge of ensuring the scientific and ethical relevance of the research project.

5.3.2 Use of a Biological Collection

5.3.2.1 Collections Regulated by the Jardé Law

Samples from new collections declared in a research protocol regulated by the Jardé Law can be used according to the conditions established in the project description.

5.3.2.2 Collections Not Regulated by the Jardé Law

These collections contain biological samples collected during a treatment not involved in a study or originated from mostly achieved research. These collections have a human origin, preserved in a Biological Resource Centre, and their use, as part of a new research project, is included in a non-interventional study with a change of purpose (requalification of samples).

5.4 Clinical Research and Biological Samples Regulation

5.4.1 Prior to the Jardé Law

The clinical research was divided in two categories: the interventional study and the non-interventional one (see Fig. 5.1). Biological sample-based research could be included in one or the other category, depending on the type of research and project.

5.4.1.1 Interventional Studies

Interventional studies corresponded to biomedical researches (approval from a Committee for the Protection of Persons and the competent authority, the French Agency for the Safety of Health Products (Agence Française de Sécurité Sanitaire des Produits de Santé ANSM)/French Health products Safety Agency (Agence Nationale de Sécurité Sanitaire des Produits de Santé or AFSSAPS), General Direction of Health (Direction Générale de la Santé or DGS), with a compulsory written consent) and concerned:

- all the studies involving an act on a human person, whether ill or not, with modification of the subject's usual care (Article L. 1121.1 of the French Public Health Code and the Huriet Law 2004-806): medication administration, facility implementation, surgery, physiology investigation, imagery examination..., to develop biological and medical knowledge.
- and the studies regarding care routine not governed by the Huriet Law, mainly diagnostic comparisons and therapeutic strategies (Article L. 1121.1 of the French Public Health Code). The simplified regulated procedures only required approval from a Committee for the Protection of Persons.

Joining the establishment of a biological collection required complementary proceedings.

In the specific situation where the study involved medication with collecting of biological resources (Huriet Law), completing a declaration form for the constitution of biological samples (FCEB 280806) was mandatory. This form required information regarding the purposes of the establishment of a collection, information on the

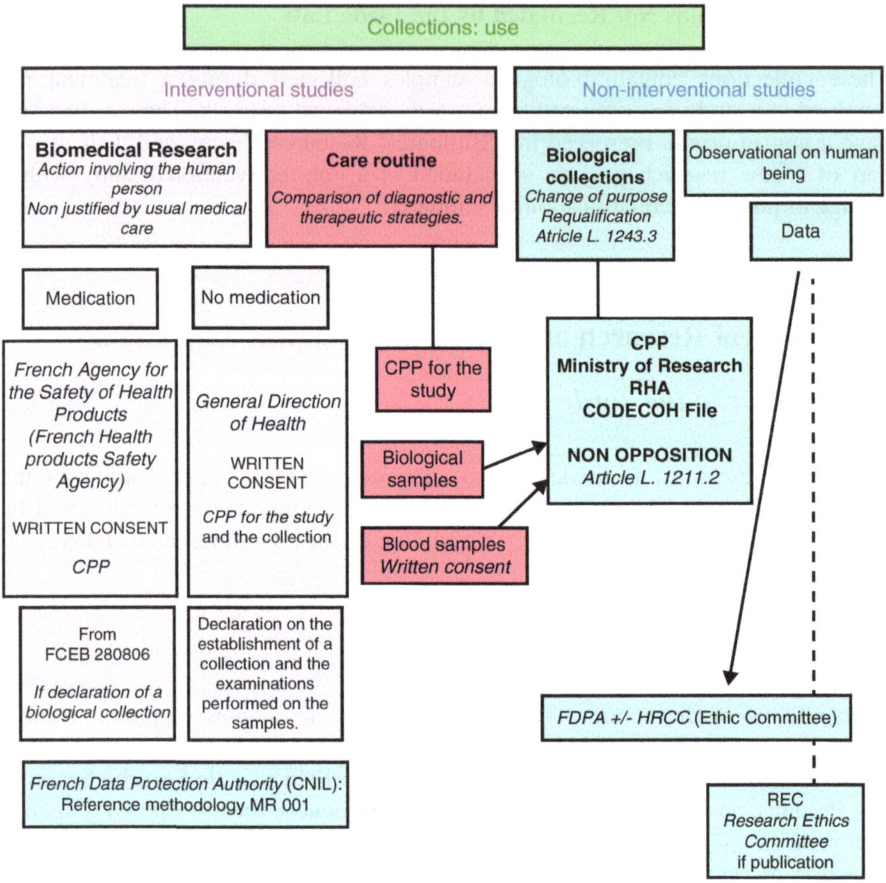

Fig. 5.1 Clinical research and biological samples regulation: prior to the Jardé Law

responsible authority, places of storage, the number and the nature of the swabs concerned. A final report and further clarifications on the future of the collection was supposed to be delivered at the end of the study: either its destruction or its use in a new research program (biomedical or not), using the legislation regulating the new study.

During a biomedical research involving no medication (Huriet Law), the collection was declared in the research protocol.

During a study regarding care routine (not regulated by Huriet Law), the use of biological specimens was submitted to a declarative regime by the organism that preserves and prepares the biological samples involving the human person (Article L. 1243-3 of the French Public Health Code). A blood test for the immediate constitution of a collection was authorized (considered as non-interventional) with the written consent of the subject.

5.4.1.2 Non-interventional Studies

The non-interventional studies combined observational studies (epidemiological studies), data analysis, and research on biological samples, whose regulation was not within the scope of the Huriet Law.

Regarding observational studies, the regulatory proceeding required an approval from the Health Research Consultation Committee or HRCC (Comité Consultatif sur le Traitement de l'Information en matière de Recherche dans le domaine de la Santé or CCTIRS), and the personal data processing was declared to the French Data Protection Authority or FDPA (Commission Nationale de l'Informatique et des Libertés or CNIL) (simple declaration). Concerning the research on data, only approval from the FDPA was required. For both types of studies, if it implied a publication, approval from a Research Ethics Committee or REC was required.

The regulation governing non-interventional studies on biological samples was based on the declaration by the organism that preserves and prepares biological samples involving the human person (Article L. 1243-3 of the French Public Health Code). For these samples already swabbed and requalified for research (change of purpose), the declaration was registered at the French Ministry of Higher Education, Research and Innovation (and at the Regional Hospitalization Agency for healthcare institutions or RHA). The files were completed in the CODECOH. In this perspective, a patient information mentioning his absence of opposition (collection of the non-opposition) for the preservation and future uses of the samples was considered as sufficient (Article L. 1211-2 of the French Public Health Code).

5.4.2 New Regulation: The Jardé Law

The application Decree for the Jardé Law (no. 2016-1537 November 16th, 2016), "relative to research involving the human person (RIHP)" was published on November 17th, 2016. It modifies the French Public Health Code on both legislative and regulatory parts. The law identifies three categories of studies, depending on the level of risk "risk-based legislation", incurred by the person volunteering for it. In these three categories, the level of regulatory constraint and the requirements for consent are different and adapted to the risk involved (see Fig. 5.2).

Henceforth, all research involving the human person (RIHP) for the development of biological and medical knowledge must receive approval from a Committee for the Protection of Persons selected at random after the deposit of the research file through a "single-window" at the General Direction of Health. The Health Research Consultation Committee disappeared. In a RIHP context, the CPP evaluates the ethical relevance of the project which will be later notified to the French Data Protection Authority (MR 0001 for interventional studies presenting minimal risks and constraints or not, MR 003 for the non-interventional studies).

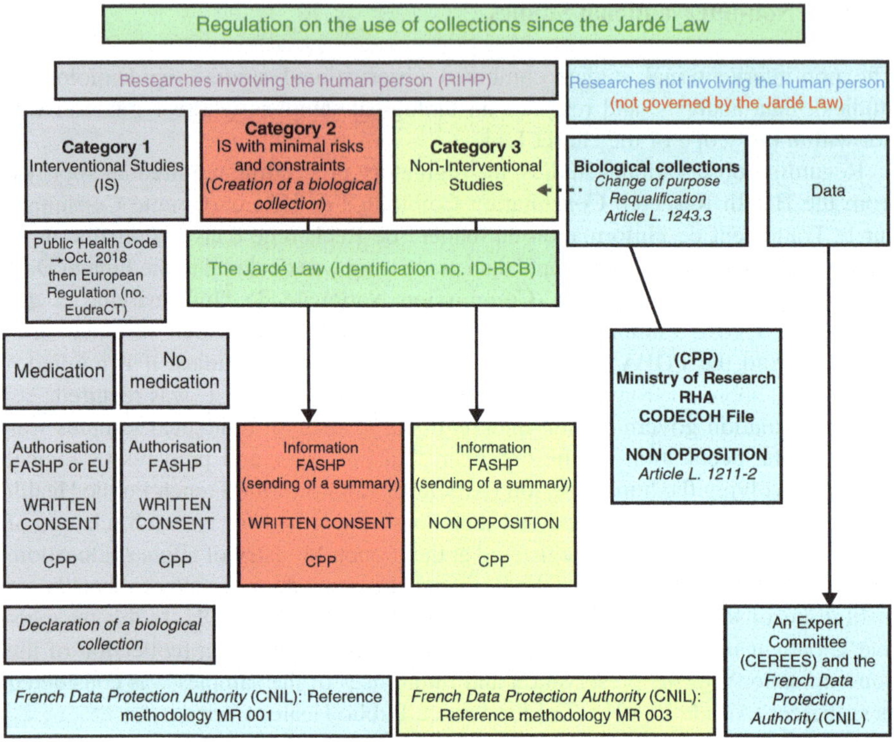

Fig. 5.2 Clinical research and biological samples regulation: the Jardé Law

5.4.2.1 Category 1 Studies: Interventional Studies Exceeding the Minimal Risks Standard (Former Biomedical Studies)

In this category, studies on medication will be governed by the European regulation from 2018 onwards. The studies require an authorization from the French National Agency for Medicines and Health Products Safety and a written free and informed consent by the subject volunteering in the study. The Clinical Research Associates must get a registration number for the research from the European Clinical Trials Database, EudraCT, (https://eudract.ema.europa.eu/) launched by the European Medicines Agency. The EudraCT number intends to identify each study conducted in one or several research areas in the European Union territory.

The procedures remain similar (prior to the Jardé Law) if the study involves a biological resources data collection. This new collection must be declared in the research protocol.

5.4.2.2 Category 2 Studies: Interventional Studies Presenting Minimal Risk and Not Involving Medication

These studies are clarified by the Decree of December 2nd, 2016: research on care routine in the former legislation, venous or arterial blood samples limited to a volume proportional to weight, swabs and biological sample collections, other than blood including skin-appendages and excreta, interviews and questionnaires whose results may conduct to a modification in the patient care. The studies in this category require notification to the French National Agency for Medicines and Health Products Safety with a summary of the study, approval from a CPP and an express (written or oral) free and informed consent.

5.4.2.3 Category 3 Studies: Non-interventional Studies Involving the Human Person and Corresponding to Purely Observational Studies

Henceforth, this category requires a simple notification to the French National Agency for Medicines and Health Products Safety with a summary of the research, approval from a CPP and the consent (non-opposition) of the patient.

Any study on human biological samples from the treatment (requalified for research), that has never received approval from a CPP, must be included in Category 3.

For these three categories of studies, except for interventional studies on medication for human use, before the deposit of an authorization application form and/or a recommendation for a research implying the human person, the Clinical Research Associates must obtain a registration number for the research on the French National Agency for Medicines and Health Products Safety website. This number intends to identify each research carried out in France. This number of registration is also called "ID-RCB number" (https://ictaxercb.ansm.sante.fr/Public/index.php). The Associates must also obtain the random nomination of a CPP and connect to the VRB application at the following address: https://vrb.sante.gouv.fr.

5.5 Function of a Biological Resource Centre (BRC): Its Collections, Its Missions: Example BRC-Cancer Toulouse University Institute for Cancer (TUIC)

5.5.1 Missions of a Biological Resource Centre

The main mission of the BRC is to ensure the management of the biological resources and collections: reception, preservation, preparation and disposal requested by a Clinical Research Associate and/or a researcher.

According to the type of studies on biological samples: constitution of a new collection involving the human person (categories 1, 2 and 3 of the Jardé Law) or use of requalified research samples not involving the human person (not regulated by the Jardé Law), the BRC guides the "applicant" to the legal department attached to its administrative structure to enforce the new regulatory proceedings.

Thus, the applicable regime for the research on biological resources with a change of purpose (former non-interventional studies on biological samples) includes the declaration by the competent authority that preserves and prepares biological samples from a human origin (Article L. 1243-3 of the French Public Health Code) at the French Ministry of Higher Education, Research and Innovation (and the Regional Hospital Agency for healthcare institutions). Files must be completed through the CODECOH and the non-opposition for the preservation and the future use of the samples is sufficient (Article L. 1211-2 of the French Public Health Code). An authorization is required for the use of collections for a cession (CODECOH). In this new legal framework, approval from the CPP is no longer required.

It is worth noting that the studies on existing data (epidemiology studies, medicine thesis on medical records) must receive the ethic approval from a new Commission, the so-called Comité d'Expertise pour les Recherches, les Études et les Évaluations dans le domaine de la Santé or CEREES; and then the study must be brought up before the French Data Protection Authority.

5.5.2 *Example: BRC-Cancer Toulouse University Institute for Cancer (TUIC)*

The BRC-Cancer of the TUIC is located at a dedicated and unique area in the Toulouse University Institute for Cancer—Oncopole. It is certified by the norm NF S96-900 guaranteeing Clinical Research Associates of high-quality biological samples. It runs independently through funding from the Directorate of Health Care Supply and gathers 18 declared and authorized emblematic collections, in accordance with French Law, at the Ministry of Higher Education and Scientific Research (DC-2008-46; AC-2013-1955) after approval from Ethical Committees (Advisory Committee on Information Processing in Material Research in the Field of Health, French Data Protection Authority, Committee for the Protection of Persons).

The 18 collections stored at the BRC-Cancer Toulouse University Institute for Cancer are all collections of tumors by organ (collection of colon cancer, lung cancer, lymphoma tumors). For a great number of tumors, a sample of healthy tissue and some serum are usually provided with tumorous tissue. The activity of receiving and preserving at the BRC-Cancer can be considered as quite elevated due to the delivery of 5500 new frozen samples per year with non-opposition from the patient regarding the use of samples for research purposes. The BRC-Cancer gathers a large amount of tumors and can provide samples for specific projects by organ or by theme: colon adenocarcinoma, glioblastoma after targeted therapies, bone metastases, rare tumors…

The main characteristic of the BRC-Cancer Toulouse University Institute for Cancer consists in being a histopathology platform dedicated to research, able to provide useful deliveries for specific research projects (production of TMAs, development of a new antibody in immunohistochemistry).

The center also offers the possibility to gather annotations regarding specific targeted projects (annotated collections).

Finally, the center has an important role at the Toulouse University Institute for Cancer arranging the disposal of samples for fundamental or clinical projects between academic collaborators of the Institute or exteriors and industrialists.

5.5.3 Modus Operandi

Once the Clinical Research Associate has contacted the BRC-Cancer through its website, and after verification and validation of the necessary steps for the provision of the biological samples, a contract is enacted between the involved parties. Different types of **contracts** (cession, delivery, collaboration) are available depending on the nature of the project. These contracts are drafted by legal counsel from the Direction of Research and Innovation at the teaching university. The cost of the different services and the medical expertise is established with a **fee schedule**, according to the involvement of the different partners in the study (type of contract, intellectual property, publication, acknowledgments), to the type of collection (scarcity, deliveries associated), and to the quality of the biological samples delivered.

5.6 Conclusion

A collection is mostly elaborated during a clinical research on human person. Its creation must provide an answer to one or several objectives around a research project to improve our biological knowledge and ensure a better care of the patient. Therefore, the collections are diversified and are either related to a specific organ pathology or as an answer to a treatment. Henceforth, these collections must respond to new regulatory requirements, intended to protect the persons involved in the project (Jardé Law).

The use of collections implying human biological samples already collected during a treatment belongs to a study that does not involve the human person, such as defined by the Jardé Law.

These collections are governed by rules ensuring the declaration and the authorization regarding the activity of preservation and preparation for human biological samples.

Collections management presumes strict organization and rigorous structuration at the BRC, submitted to certification (certification NF S96-900), to deliver high-quality biological resources and address the new challenges of biological research, both clinical and fundamental.

Thus, the BRCs plays a strategic role in the development of the research on a regional, national and international scale.

References

Lettre de la délégation à la recherche clinique d'Ile d.

MINISTÈRE DES AFFAIRES SOCIALES ET DE LA SANTÉ Décret n° 2016-1537 du 16 novembre 2016 relatif aux recherches impliquant la personne humaine, NOR : AFSP1621392D.

Chapter 6
Does Tumor Bank Regulation Promote the Valorization of Tumor Collections?

Anna Pigeon

Soon enough, the legislature has become aware of the scientific and economic value of tumor collections, and more generally of biological sample collections, and has progressively regulated their use (Bellivier and Noiville 2009; Rial-Sebbag 2009). First, the therapeutic and scientific use of detached elements of the human body[1] has been legitimized. Their use is subject to the respect of the principle of dignity of the human person, from which stems the consent to collection and donation, anonymity and non-patrimony.[2] Second, the processing of personal data for research purposes in the field of health[3] has also been regulated. The scientific value of the samples being closely related to their association with personal health data, and the creation of a legal regime that is conducive to the use of this data, has facilitated the constitution of biological sample collections and encouraged increasing research carried out from those resources.[4] Finally, the activity of the agencies in charge of collecting these biological samples, preparing them, organizing them, storing them and making them

[1]Law No. 94-653 of 29th July 1994 relating to the respect of the human body and law No. 94-654 on the donation and use of elements and products of the human body.

[2]From Article 16 of the Civil Code.

[3]Law No. 94-548 of 1st July 1994 related to the processing of personal data for research purposes in the field of health and amending law No. 78-17 of 6th January 1978 "Informatique, Fichiers et Libertés", French Data Protection Act.

[4]This system notably includes the lifting of medical secrecy necessary for the transfer of data for health research purposes, as well as that of no objection (implicit consent) to the processing of such data for research purposes in the field of health.

A. Pigeon (✉)
University Toulouse 1 Capitole, Maurice Hauriou Institute, Toulouse, France

Tumor Banks Project, Toulouse, France

X. Bioy (ed.), *Public Regulation of Tumor Banks*,
https://doi.org/10.1007/978-3-319-90563-1_6

available to researchers, has been regulated.[5] The latter must, according to their activities, meet the administrative procedures of declaration or authorization.[6]

In 2004, the legislature became aware that biological samples must be developed and cannot stay unused.[7] This increased awareness has materialized with the adoption of legal rules facilitating the re-use of biological samples (Thouvenin 2005a, b). On the one hand, samples can be re-used for purposes other than those for which they were collected, people only having the right to object after having been duly informed,[8] and there is a wider use of elements which may be used for scientific research.[9] The re-use of personal data for research purposes is also subject to the principle of presumed consent after having been informed. On the other hand, exceptions to the obligation to provide information are implemented regarding samples[10] and data,[11] when it is impossible to find the persons concerned.

In recent years, a true "public policy of BioBanks " (Bioy 2010) has also emerged to promote their development. A growing policy of provision of biological samples for research has been set up under INCA supervisory thanks to "funding, an annual follow-up of the activities and the development of recommendations to facilitate

---

[5]Law no. 96-452 of 28th May 1996 on various health, social and statutory measures; Law no. 98-535 of July 1st, 1998 strengthening the health monitoring and control of products for human safety; Law No. 2004-800 of 6th August 2004 relating to bioethics and its successive reviews.

[6]Article L. 1243-3 and L. 1243-4 of CSP.

[7]In reference to the objectives set by Article L.111-1 of the Research Code:

> the national policy of research and technological development aims to:
>
> (1) increase knowledge;
> (2) share scientific, technical and industrial culture;
> (3) develop the results of research to serve society[...]

[8]Article L. 1211-2 of the Public Health Code.

[9]Article L. 1211-2 of the Public Health Code.

[10]Article l; 1211-2 of the Public Health Code: " (. . .) It is possible to bypass the duty to provide information when it is impossible to find the person concerned, or if one of the Advisory Committees for the protection of persons referred to in article l. 1123-1, consulted by the research leader, does not consider this information to be necessary (...)".

[11]"Article 57 of the Data Protection Act:" II. -When personal data has been initially collected for a reason other than research, study or assessment, it can be derogated, subject to III, from the information obligation defined in I:

(1) for the process required for the conservation of this data for historical, statistical or scientific purposes, under the conditions provided for in book II of the Heritage Code;
(2) when individual information does not enable to find individuals.

(11 cont.) The requests for exemptions to the obligation to inform individuals of the use of their data for research, study or assessment purposes are justified in the approval file sent to the National Commission for ICT and freedoms, which will make a final decision.

their activities".[12] Biobanks are also encouraged to be certified; the allocation of funding is being conditioned to harmonize their practices in terms of structure management and biological resource quality.[13] Finally, they have organized into specialized and excellence networks at a local, national, European and international level, which has allowed them to increase their visibility in terms of standardizing consents or of participating in the building-up of large cohorts (Washetine et al. 2017).

However, the regulation of biobanks has gradually developed following scientific and normative needs, which has led to the coexistence of a heterogeneous set of rules. Several regulations are applicable depending on the context of how the following were collected: elements of the human body and associated personal data, their conditions of conservation, and the purpose of their use. Thus, tumor banks must combine the provisions of Bioethics law, Public health laws with those of the Data Protection Act,[14] and with that of Research involving human individuals.[15] These regulations are complemented by a corpus of non-binding standards set by states or non-governmental international forums,[16] scientific societies or ethics committees. The accumulation of rules can mean hindrance and difficulties for those concerned. Therefore, the question that arises is the following: do biobank regulations, consisting of scattered legal rules, constitute a brake on the development of these activities, and do they promote the development of the biological sample collections?

However, we need to briefly clarify what is meant by valorization and its purpose in our study. The term valorization is used in different branches of law. Yet, the legislature has never defined it, and it can have different meanings depending on the object to which it relates.[17] In our study, we shall refer to the *legal vocabulary* (Cornu 2016) that defines valorization as "the act of enhancing the intrinsic value of a thing by work, (. . .)". The idea is to develop these materials so that they become resources for other actors.[18] It comes to valorizing, tumors collected under medical

---

[12]Institut National du Cancer, Scientific report 2015/2016, 2016, p. 93.

[13]NF S96-900 September 2011, Quality of biological resources centers (CRB)—Management System of a CRB and quality of biological resources. At an international scale: Normes ISO 9001.

[14]French Data Protection Act: Law no. 78-17 of 6th January 1978 (" Informatique, Fichiers et Libertés").

[15]Law no. 2012-300 of 5th March 2012 relating to research involving human individuals.

[16]OECD, Guidelines for Human Biobanks and Genetic Research Databases, October 22nd 2009.

[17]The French term "valorisation" (here, "development" in English) can mean assessing the value of an object or increasing its value.

[18]Valorization may however include an economic dimension being subject to other contributions—see, in particular, Carole Haritchabalet and Catherine Bobtcheff -, we shall focus on developing scientific collections. That being said, we must reiterate that development is conducted here on behalf of the public service for research, and on that basis, it must contribute to satisfy general interest. Thus, economic development cannot have other purposes than the maintenance and improvement of the public service for research, its goal being to meet the needs of the population in terms of health.

care in the interest of the patient, into a resource for public or private researchers. Thus, the immediate cause for valorization is scientific research. Resources cannot be hoarded, they must circulate so that research can be conducted.

Concerning the object of valorization, it focuses on the collections of biological samples which include four different legal objects, each having their own judicial regime: biological samples, associated personal health data, the expertise of the collector and the database. This set is assigned to scientific research. And, as a whole set, it must circulate so that research can be conducted. The former should be valorized, in the interest of science and society in general. However, collections are the object of development which biobanks are responsible for. They are infrastructures (Rial-Sebbag 2008), the superior level of organization of which enables the circulation of samples. Our comments will focus on biobanks aimed at research, and more particularly on tumor banks. However, the organization of these structures, the means they have, and the legal rules they must respect, are predominant factors in the valorization of collections.

The coordination of different regulations that correspond to the protection of various interests is not exceptional,[19] but in biobanks, it generates many difficulties.[20] The availability of biological samples to researchers, and so their scientific development, can thus be hampered by a too complex legislation which is difficult to access for its recipients. Aware of the complexity of the legal regime of biobanks, and of the importance of collections development, the legislature has decided to simplify the regulation. This simplification is to do with the management of biobanks activity (Sect. 6.1) and with access to associated data (Sect. 6.2). Yet, despite this political will for simplification, its effects are limited in practice.

6.1 What Does a Simplification of the Regulation of Biobanks Activity Entail for Their Valorization?

The scientific valorization of collections involves an improvement of their accessibility and their circulation. To do this, the legislature has wanted to simplify the management of biobanks activity for specific research purposes (Sect. 6.1.1), as well as the constitution of biological sample collections (Sect. 6.1.2).

[19] A number of activities must respect various regulations.

[20] See on this point: Assemblée Nationale, Presentation of reasons for the Bill on research on the human person, recorded at the Presidency of Assemblée Nationale on January 6th, 2009, presented by Mr. Jardé.

6.1.1 Simplifying the Management of Biobanks Activity for Specific Research Purposes

Law No. 2012-300 of March 5th, 2012, relating to research involving the human person,[21] has considered the need to develop the activity of biobanks and promote the circulation of collections. To do this, it simplifies procedures by distinguishing the supervision of activities of preparation and conservation of samples on the one hand, and that of research conducted on the collections on the other hand.[22]

The judicial regime of collections based on the 2004 bioethics laws and public health laws was indeed difficult to implement. Article l. 1243-4 of the Public Health Code placed the process of preparation and conservation of samples for specific research purposes, and the creation of collections, under the control of the Ministry of research and the Agence Régionale de la Santé (ARS, "Health Regional Agency") which deals with health structures, and a Research Ethics Committee (In French, "Comité de Protection des Personnes" or CPP). The first two instances proceed to the verification of compliance with regulatory requirements, in the form of opposition or suspension of activities. Meanwhile, the CPP would be called upon under Article l. 1243-3 of the CSP[23]and assess "the quality of the information given to participants, the terms of consent and the ethical and scientific relevance of the project". Yet, this system stemming from the unification of the two regimes created by the 1996[24] laws framing the aim of genetic research collections, and by the 1998[25] law governing the activity of the agencies preserving samples for scientific purposes, "has led to a genuine deadlock of sample collection activities[26]". Building a collection already implies a research operation—selection and collection of samples for research purposes, according to the clinical characteristics of one or more members of the group—and the activity of the biobank can be summed up in the preparation and conservation of samples stemming from medical care. The intervention of the CPP is then questionable.

[21]Law passed in 2012, finally applied 4 years later after the publication of Decree No. 2016 - 1537 of November 16th, 2016 relating to Research involving the human person.

[22]Assemblée Nationale, Report no. 1377, on behalf of the Cultural, Family and Social Affairs Committee, on Mr. Olivier Jardé's bill (no. 1372) relating to research on the human person, January 13th, 2009.

[23]Article l. 1243-3 of the Public Health Code: "(...) The organizations mentioned in paragraph 1 submit their draft statement to the prior notice of a committee, as defined in chapter III of title II of book I in this part, the mission of which is to assess the quality of the information given to participants, the terms of consent and ethical and scientific relevance of the project. The statement is addressed to the Minister in charge of research and, where applicable, to the relevant Director general of the regional health agency, concomitantly to the submission for opinion to the Committee. The latter's view is given to them without delay (...)."

[24]Law no. 96-452 of 28th May 1996 on various health, social and statutory measures.

[25]Law no. 98-535 of July 1st, 1998 on strengthening the health monitoring and control of the safety of products intended for humans.

[26]Assemblée Nationale, Report no. 1377.

How can the latter evaluate ethical and scientific relevance of a preparation and conservation activity that generally contains samples which have not been clearly assigned to a specific research project? Is it about assessing the general research subject of an organization, such as cancer for example, or the subject of each collection? In the first hypothesis, how could the CPP evaluate "the quality of the information given to participants, the terms of consent and the ethical and scientific relevance of the project" which, most of the time, has not been defined yet?

In practice, some CPPs have even refused to issue an opinion, which has hindered several research projects using biological sample collections.[27]

Therefore, the law has simplified the regulation and separated the two activities. The declaration of activity of a biobank is now only subject to the Minister of the research, and not to the CPP.[28] This new provision, which is more in line with the activities carried out,[29] allows the biobank manager to reduce the time spent on administrative formalities, which facilitates his activity and his mission of provision of samples.

For projects of biological sample collections, the simplification is far from obvious.

6.1.2 The Simplification of Procedures Regarding the Build-Up of Biological Sample Collections

The project of biological sample collection is now governed by the legal framework of research involving the human person,[30] which distinguishes three categories of research (Chemtob-Concé and Cailleux 2013) according to the risks and constraints[31] experienced during research (Dionisi-Peyrusse 2012). It should be remembered that the main contribution of the law called "Jardé", lies in the creation of a common framework for all types of research (Galloux and Gaumont-Prat 2017)

[27] Assemblée Nationale, Report no. 1377.

[28] Article L. 1243-3 of the Public Health Code.

[29] The government's research department monitors respect for safety, the conditions in which no objection can take place and the subjects of the preserved collections.

[30] Law no. 2012-300 of 5th March 2012 on research involving the human person.

[31] The first category includes interventional research and corresponds to current biomedical research, for which the regime remains unchanged. A second category includes "Interventional research which does not involve drugs, with minimal constraints and risks, the list of which is decided by order of the Minister of health, on the advice of the Director general of the national agency for medicine and healthcare products". This new category is close to (33 cont.) the former category of "research to assess routine care", except that its scope is wider since this category includes interventions without risk. Finally, the third category, one in which all acts are performed and products are used in the usual way, without any additional or unusual diagnosis, treatment or monitoring procedure, corresponds to the former category of non-interventional research. A new feature requires the opinion of the CPP for any research.

based on the mandatory opinion of the CPP (Thouvenin 2012). For collections management, the major benefit of this law lies in the fact that the collections do not have to be presented to the Ministry in charge of research and, where appropriate, to the regional health agency. However, it is up to the developer to qualify their research under the three types of research provided by the law. But this operation is so complex that the law itself provides in Article L.1121-4 of the CSP, "that in the case of serious doubt about the qualification of research under the three categories involving the human person defined in Article l. 1121-1, the committee for the protection of the people concerned takes advice from the national agency of drug and health product safety". Thus, collections can be subject to three different legal regimes.

When collections are made in support of interventional research, they are subject to the legal regime applicable in 1 ° of Article l. 1121-1 of the CSP. They must, at the same time as the rest of the protocol, receive a favorable opinion from the CPP and be authorized by the National Agency for medicine ("ANSM" in French, "l'Agence Nationale de Sécurité du Médicament"). Prior to submitting their authorization application and/or opinion request application on research involving the human person, the proponents must also obtain a registration number from their research on the website of the ANSM (no. ID—RCB).[32] This number is meant to identify each research project conducted in France. It is only at the end of a research project that an activity statement will be submitted to the Ministry of research, if the collection is preserved at the end of the research project.[33]

On the other hand, there are still uncertainties about the legal regime to be applied to biological sample collections built-up outside interventional research, having been built-up using re-classified samples (samples from medical care or previous research), or using samples specifically collected to be part of a collection.

Can they be considered as low-grade interventional or non-interventional research? The Decree of May 3rd, 2017 giving the list of research mentioned in 2 ° of Article l. 1121-1 of the Public Health Code, provides that the sampling and collection of biological resources, other than blood, specifically for research purposes (the number, volume and/or the size of the biological samples collected are justified in the research protocol), are interventions with only minimal constraints and risks. Because of the absence of debate on this issue, we can assume that the notion of collecting, here, refers to samples that have already been collected and which have been reclassified. As for the notion of sampling, it could refer to samples taken directly to build up a collection. This would mean providing a legal framework

[32]This number is used to identify all research carried out in France.

[33]It will only be complemented with a new notice from the Research Ethics Committee if the built-up sample collection is used for a new research project, different from that for which it was originally built-up. "Therefore, there will not be two separate processes anymore, one to perform and report on research, the other to build up the biological collection".

for sampling which would directly be for building up collections, which until now had existed for blood[34] samples, but has been dropped.

In both cases, the collections are subject to the regime applicable to research with minimal risks and constraints.[35] Under Article l. 1121-1 of the CSP, they may be implemented only after a favorable opinion from the Committee for the protection of people and after having informed the ANSM.[36]

However, the collections built up outside research involving human persons remain subject to Articles l. 1243-3 and 1243-4 of the CSP: prior statement or approval from the Minister of research and the ARS for healthcare facilities. The statement will be supplemented with a new notice from the Committee of protection of individuals in cases where the built-up collection of samples would be the subject of a new research project, different from the one for which it had been originally designed.[37]

The complexity of the new system, the multiplication of specific provisions and uncertainties around the qualification of research does not constitute a simplification of procedures for the build-up of collections and *a fortiori* for their subsequent circulation. The political objective might have been to develop the collections through making them available to researchers, but the actual effects are very limited. This observation is even more damaging that these provisions must be associated with those relating to data protection, an area where simplification is also very relative.

6.2 Are We Moving Towards a Simplification of Data Access?

Biological samples prepared and stored in tumor banks are associated with data related to donors, including clinical, biological, or even epidemiological data. Access to increasingly full-on data is essential to put together biological samples collections of great scientific value.[38]

To valorize health data, law No. 2016-41 January 26th, 2016, on the modernization of our health care system, called "Touraine law", has reformed procedures in

[34]Article L. 1221-8 of the Public Health Code, as amended by order No. 2016-1406 October 20th, 2016 which removes the reference to biological sample collections.

[35]Decree of May 3rd, 2017, giving the list of the research referred to at 2 ° of Article l. 1121-1 of the Public Health Code.

[36]The proponent gives a copy of the notice of the CPP and a summary of the research to the competent authority.

[37]Articles L. 1123-7 and L. 1123-9 of the Public Health Code and R. 1123-22 of the Public Health Code.

[38]Collection managers should endeavor to collect maximum information from patients and increase links between sample data and the patients at the origin of these samples. See on this point: Institute National du Cancer, Scientific Report 2015/2016, 2016, p. 93.

terms of access to health data (Sect. 6.2.1). In biobanks, the contributions of this law, however, are limited (Sect. 6.2.2).

6.2.1 The Simplification of the Implementation of Data Processing Procedures

Health data[39] collected by tumor banks is sensitive personal data under Article 9 of Regulation No. 2016/679[40] and Article 8 of the French Data Protection Act. Their use for health research purposes is subject to the general provisions of the French Data Protection Act as well as the specific provisions of chapter IX, relating to the processing of personal data for medical research. Tumor banks must respect the rights of the persons concerned; their data can be collected as soon as the duly informed person has not objected,[41] and they can submit their activities to a control carried out by the French Data Protection Authority (CNIL) which rules in the form of an authorization.[42]

Meanwhile, the processing of personal data implemented as part of a project of biological sample collection must be authorized by the CNIL, based on the public interest in the research, and after the notice of a Research Ethics Committee (CPP).[43] This new procedure is the result of a change of chapter IX of the Data Protection Act, through the Law of modernization of our healthcare system. The objective here is to simplify provisions deemed too complex to facilitate open access to health data. Under the previous system, the processing of personal data for health research had to be authorized by the CNIL on the opinion of the Advisory Committee on Research Information Processing in the field of health (referred to as "CCTIRS" in French).[44] However, this advice would overlap the CNIL authorization, and, based on the framework governing the collection, that of the CPP, to which should be added a statement to the Department of research and authorization from the ANSM. The CCTIRS has now been dropped and replaced by the CPP that is, in addition to its usual functions, in charge of assessing the risks associated with the protection of personal data. However, this procedure is simplified as long as the process is

[39]Article 4, § 15 of the regulation on the protection of personal data: "health data, personal data relating to physical or mental health of an individual, including the provision of healthcare services which reveals information on the health of this person".

[40]Regulation (EU) 2016/679 of the European Parliament and of the Council of April 27th, 2016, relating to the protection of natural persons with respect to the processing of personal data and the free movement of such data, and repealing directive 95/46/EC (general regulation on data protection).

[41]Article 56 al. 1 of the Data Protection Act.

[42]Chapter IV of the Data Protection Act.

[43]Article 54 of the Data Protection Act.

[44]Article 54 of the Data Protection Act.

implemented as part of the referring methodologies MR001[45] and MR003[46] recently published by the CNIL. In both cases, a simple commitment to compliance with the referring methodologies is required. Finally, for research not involving human persons, or in other words for research on data,[47] the CNIL must issue an authorization after the advice of an Expert committee for research, studies and assessments in the field of health (referred to as "CEREES" in French).[48]

This system seems to simplify the preliminary procedures governing data processing, and thus the build-up of biological sample collections, but, in practice, it turns out to be complex, and the simplifications are not very effective.

6.2.2 *Is It an Effective Simplification?*

To assess the effectiveness of the standard in terms of valorization, one should wonder about the elaboration of this standard and its objective. We have seen that the objective pursued by the law of modernization of our healthcare system is to create the conditions for open access to health data.[49] And here lies the idea governing the valorization of biological samples: patients' data is transformed into resources for researchers; while protecting confidential data, the latter must be developed, i.e. be available and circulate so that scientific research can be conducted. To this end, the law reforms and clarifies the authorization system of personal data processing for research, study and assessment purposes in the field of health.[50] Yet, the clarification seems far from obvious, and for tumor banks its effects seem limited. On the one

---

[45]Deliberation No. 2016-262 July 21, 2016, amending the referring methodology of the processing of personal data used in biomedical research (MR-001).

[46]Deliberation No. 2016-263 July 21st, 2016, which approves of a referring methodology relative to the processing of personal data implemented as part of research in the field of health, which does not require explicit or written consent from the person concerned (MR-003).

[47]Nowadays, a part of the activity of tumor banks is to create clinico-biological databases (referred to as "BCB" in French), the funding of which is on the increase. These are "tools of collection and data management relating to individuals, patients and subjects involved in research protocols. Data associates clinical information, scientific and medical tests performed on biological samples, the epidemiological information and administrative information collected from surveys or questionnaires". The research carried out using these tools focuses on data and must be authorized by the CNIL after being assessed by the CEREES.

[48]In this respect, one can wonder about the relevance of the creation of a Committee that has more or less the same missions as the former CCTIRS. Article 54 of the LIL states that the latter "issues, within a period of 1 month from referral, advice on methodology, on the need for the use of personal data, on its relevance regarding the purpose of such processing and, if it is necessary, on the scientific quality of the project." The same applies to its composition which is almost identical.

[49]National Assembly, report No. 2302, Bill relating to health, presented by Ms. Marisol Touraine, Minister of Social Affairs, health and women's rights, in the name of Mr. Manuel Valls, Prime Minister October 15th, 2014.

[50]Chapter IX of the Data Protection Act.

hand, several separate legal regimes are applicable depending on the category of research involving the human person in which processing happens, and the systematic and exemption regimes should work together (MR001 and MR003). The collection manager must qualify his research within the framework in which processing is performed but also according to its purpose, which is a complex operation.

On the other hand, Article 54 of the LIL now states that the CNIL should allow processing based on the public interest in the research, this interest having to be demonstrated and justified, which is challenging for private biobanks. In this respect, the law provides that "in the conditions set by decree in a Council of State, the National Institute of health data provided for in Article l. 1462-1 of the Public Health Code, can be submitted to the National Commission on ICT and freedoms or to the Minister of health regarding the public interest of the research, the study or the assessment justifying the claim for processing; it can also raise the case on its own initiative".

Finally, authorization requests are now deposited with the entrusted single secretariat, according to 2 ° of Article l. 1462-1 of the Public Health Code, at the National Institute of health data (referred to as "INDS" in French), which forwards them within a maximum period of seven working days to the appropriate Committee to deliver an opinion on the project.[51] Does the intervention of an additional public authority, added to the already substantial number of competent authorities in this area, constitute a simplification of procedures?

Furthermore, one can wonder about the new competence of the CPP on data protection. It should be remembered that the latter replaces the CCTIRS for data processing implemented as part of research involving human persons.[52] Of course, this replacement lightens procedures since it is not necessary to seek the views of the CCTIRS anymore. But do the CPPs have the material and substantial capacity to absorb its competence?

The mission of the CCTIRS,[53] consisting of "competent people in health research, epidemiology, genetics and biostatistics" was to enlighten the CNIL on the justification of processing. It had been given a role of scientific expertise by the legislature. The idea is that it is for peers to judge the scientific validity of a research project (Coulibaly 2011), and for that only scientifically competent persons can

[51] Article 20 of Decree No. 2016-1872 of December 26th, 2016, amending Decree No. 2005-1309 of 20th October 2005 taken for the application of law No. 78-17 of 6th January 1978 relative to information technology, files and freedoms.

[52] Article 54 of the LIL.

[53] CCTIRS consisted in "competent people in health research, epidemiology, genetics and biostatistics" and had to deliver their opinion on "the research methodology in the light of the provisions of this law, the need for the use of personal data and the relevance of these compared to the research objective".

intervene.[54] Are the CPPs scientifically competent for this type of analysis and do they have the necessary means to absorb the inevitable increase in requests?

The complexity of the rules, the addition of committees, and the multiplication of standards does not constitute a simplification of the legislation yet necessary for the effectiveness of rules in this area. One can see that it will take time to promote the provision of biological sample collections for researchers, and strike the right balance between protection of individuals, health security, and the scientific development of collections.

References

Bellivier, F., & Noiville, C. (2009). *Les biobanques* (128 p). PUF, coll. Que sais-je?.

Bioy, X. (2010). Vers une politique publique des biobanques. *Revue de droit sanitaire et social*, 885.

Chemtob-Concé, M.-C., & Cailleux, A. (2013). L'impact des nouvelles dispositions de la loi relative aux recherches impliquant la personne humaine. *Médecine et Droit*, 30–35.

Cornu, G. (2016). *Vocabulaire juridique* (11th ed.). Paris: Puf. coll. Quadrige.

Coulibaly, I. (2011). *La protection des données à caractère personnel dans la recherche scientifique.* PhD, Université de Grenoble, 1117 p., p. 156.

Dionisi-Peyrusse, A. (2012). The law of March 5th, 2012 on research involving the human person: looking for a balance between protection of the person and development of research for medical progress. *Revue juridique personne et famille.*

Galloux, J. C., & Gaumont-Prat, H. (2017). Droits et libertés corporelles, February 2016–February 2017. *Recueil Dalloz.*

Rial-Sebbag, E. (2009). *Implication juridiques des nouvelles formes de gouvernance en biotechnologies : l'exemple des biobanques utilisées en recherche.* PhD, Université Toulouse Paul Sabatier.

Rial-Sebbag, E. (2008). Genèse d'un cadre réglementaire pour les collections d'échantillons biologiques utilisées en recherche. Exploration d'un modèle de gouvernance. *RGDM, 27*, 233–271.

Thouvenin, D. (2012). La loi n°2012-300 du 05 mars 2012; des recherches pratiquées *sur* la personne aux recherches *avec* la personne. *RDSS*, 787.

Thouvenin, D. (2005a). Les banques de tissus et d'organes : les mots pour le dire, les règles pour les organiser. *LPA*, (35) 31.

Thouvenin, D. (2005b). La loi relative à la bioéthique ou comment accroître l'accès aux éléments biologiques d'origine humaine. *Recueil Dalloz*, (2), 116–121.

Washetine, K., Ilié, M., Hofman, V., Butori, C., Long, E., Lassalle, S., et al. (2017, 2017). Les biobanques, quels enjeux en 2017. *Revue francophone des laboratoires*, (488), 25–27.

[54]Depending on that opinion, the CNIL may or may not pronounce on the need to waive professional secrecy that protects personal health data and allows the proposed data processing.

Chapter 7
Economic Modeling and Valorization of Biobanks

Catherine Bobtcheff and Carole Haritchabalet

Biobanks are service-provider facilities that collect and store biological resources which can then be made available to academic and commercial researchers. These resources are an essential component of medical research, as they enable researchers to test hypotheses and develop innovations aimed at improving the diagnosis and therapeutic management of patients. Hence, a biobank's purpose is to produce high-quality biological resources, a process that requires combining and coordinating a wide range of scientific and technical expertise. In addition, biobanks must comply with numerous legal and regulatory requirements, particularly with respect to sample collection, sample transfer, and the management of personal data. Collecting this data, which includes all the information relating to a sample, is a difficult but vital part of a biobank's work, as it enables researchers to make the best possible use of each sample.

Hence, the economic valorization of biobanks is mainly a question of valorizing research. Successful research is dependent on access to high-quality samples, provided by biobanks, and on the degree of involvement of the actors in the sample production chain. However, the provision of samples can be impacted by information asymmetries in the biobank-research team relationship. The economics literature contains numerous studies analyzing the effect of information asymmetries on the economic valorization of inputs in the innovation process. A key issue in this respect is ensuring contracts provide the parties with adequate remuneration for the effort they provide, while maintaining incentives for innovation (Aghion and Tirole

C. Bobtcheff (✉)
Toulouse School of Economics, CNRS, University of Toulouse Capitole, Toulouse, France
e-mail: catherine.bobtcheff@tse-fr.eu

C. Haritchabalet
University of Pau and Pays de l'Adour (CATT), Pau, France

Toulouse School of Economics, Toulouse, France
e-mail: Carole.Haritchabalet@tse-fr.eu

X. Bioy (ed.), *Public Regulation of Tumor Banks*,
https://doi.org/10.1007/978-3-319-90563-1_7

1994; Tirole 1999). Contracts can achieve this by sharing innovation property rights between the different parties and by including license and royalty payments.

Nevertheless, the specificities of biobanking demand a more focused economic analysis. Biobanks' decisions, especially those concerning its strategic positioning, dictate how it fits into the national and international biobanking landscape, and have an impact on both market structure and performance and therefore on the market for biological samples. They may also affect the way a biobank operates. Decisions about the size and scope of its collections determine the biobank's nature as a generalist or specialist facility. Another decision that can have a major impact on the success of research projects is whether to work with other biobanks within a network.

We describe here the economic functioning of biobanks and identify the main elements affecting their profitability, and therefore viability, and the dynamics of the innovation process. Our proposed framework provides biobanks with a tool they can use to analyze the economic effects of their decisions.

7.1 Economic Operation of a Biobank

How biobanks are structured and run varies from one facility to another, but several characteristics are fundamental to the production and supply of samples. Figure 7.1 summarizes the way biobanks operate and highlights the various elements involved in their economic valorization.

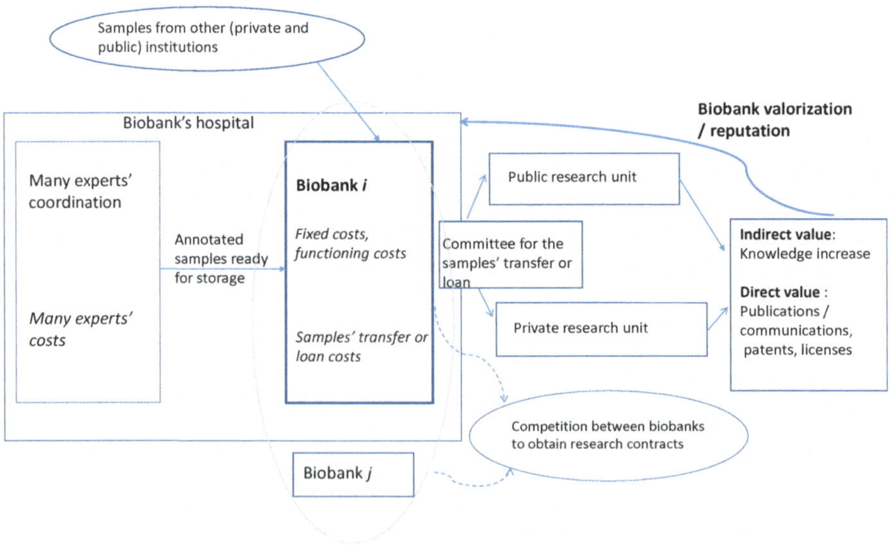

Fig. 7.1 Economic functioning of a biobank

A biobank is usually attached to a hospital, from whose patients it obtains the vast majority of its samples. Subject to patient consent, any residues of samples taken for therapeutic purposes, together with all corresponding data, can be stored in the biobank. In addition, some biobanks supplement their collections by obtaining samples from other institutions. Collecting and storing samples entails fixed and variable operating costs (depending on the number of samples stored). A large proportion of these costs are related to the work of coordinating the different professions involved in building collections.

Research units, whether private or public, contact biobanks to obtain the samples they need for their research, which may be fundamental or applied. Research projects may be conducted independently or in conjunction with the biobank. A research unit can only acquire samples if the biobank agrees to supply them, a decision that is made by an access committee composed of scientific experts and hospital managers. These decisions are based primarily on the research project's scientific quality and scope, but they may also be influenced by the public or private nature of the research unit. Some samples are transferred permanently, in which case the biobank gives up all control over their use. Other samples are lent to the research unit, thus paving the way for scientific collaboration with the supplying biobank.

If the biobank agrees to supply samples, the biobank and research unit draw up a contract stipulating each party's rights and obligations. The terms of this contract can have a major impact on both the provision of samples and the success of the research project. One of the contract's most important clauses is the tariff at which samples are transferred, which may vary according to whether the research unit is public or private. Other clauses may cover property rights, licenses, and royalties. The monies a biobank receives under its contracts with research units must at least cover its operating costs, although the sums exchanged for samples depend on the bargaining powers of the actors and on the amount of competition between different biobanks for the same project.

In addition to the financial aspects noted above, the economic valorization of biobanks is intimately linked to the success of the research projects with which they are associated. Research projects are a form of innovation focused on knowledge creation and for which it is impossible to predict the outcome. Because the impact of research projects can be evaluated in academic terms, as well as in monetary terms, patents, licenses, and publications are key indicators in assessing a biobank's quality and reputation.

Exchanges between biobanks and research units must be considered within a market framework. Biobanks may have one or more competitors, either nationally or internationally, depending on the type of sample a research unit requires. This raises the issue of whether it is advantageous for biobanks to work together in a network. In addition to reducing operating costs and expanding the range or quantity of samples on offer, networking can increase a biobank's bargaining power. However, it may also require biobanks to accept a certain loss of control over their collections and result in them losing their specificity and reputation.

This brief analysis of the economic functioning of biobanks enabled us to identify the challenges involved in valorizing biobanks, which we discuss in the following section.

7.2 Challenges Affecting Valorization

7.2.1 Information Asymmetries

Sample quality has a major impact on a research project's chances of success. However, information asymmetries may make it difficult for research units to assess the quality of available samples and thereby restrict the possibilities for entering into exchanges with biobanks. In economics, there are two types of information asymmetry. *Adverse selection* occurs when one party holds private information that is not observable by the other party. *Moral hazard* occurs when one of the parties could perform actions (which we refer to as "*effort*") that are not observable by the other party. Figure 7.2 shows potential sources of information asymmetry in a biobank's operations.

Adverse selection is a potential problem for both biobanks and research units. For biobanks, adverse selection issues arise from the difficulty of assessing either a research unit's expertise or a project's chances of success and potential scientific impact. Biobanks try to overcome this information asymmetry by setting up access committees to identify the most promising research projects. In the case of research units, adverse selection's problems affect the choice of biobank to work with, as it is

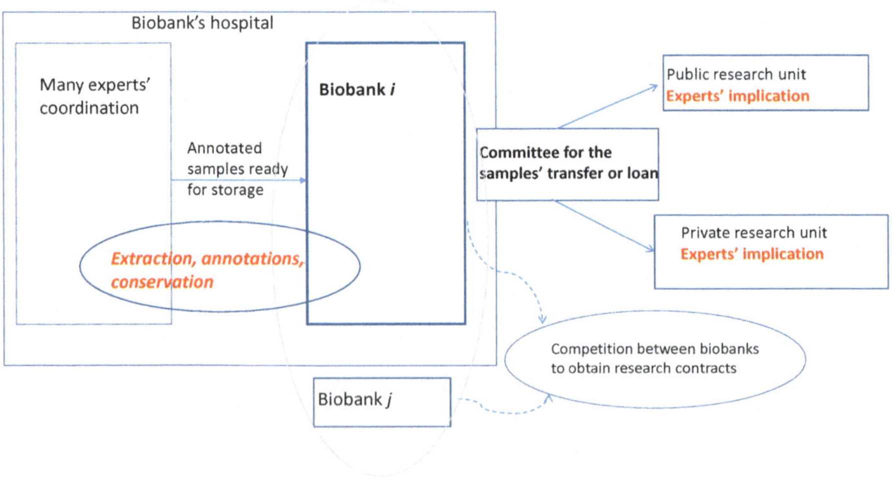

Fig. 7.2 Information asymmetries in the functioning of biobanks

difficult to make *a priori* assessments of sample quality. Consequently, biobanks must signal their quality, often by implementing costly actions such as obtaining certification or accreditation. Developing long-term relationships with research units and/or being able to demonstrate a proven academic record mitigates this problem for more experienced biobanks.

The most common form of information asymmetry biobank managers must deal with is moral hazard. Moral hazard can arise in two areas of biobank's operations: the "extraction, annotation, and conservation" process, and the "implication" stage. The care with which the people responsible for carrying out and coordinating the different stages in the extraction, annotation, and conservation process can vary, and this can lead to significant variation in sample quality. A moral hazard problem may also arise if biobank's experts continue their involvement in the project after the samples are transferred, by, for example, providing scientific support to the research unit. In such cases, a biobank can continue to manipulate the quality of the samples after the contract is signed. These moral-hazard problems are most frequently addressed by making payments to the biobank conditional on the research project's outcome. Conversely, the biobank is awarded in a way that encourages it to do as much as it can to ensure the project is a success (Aghion and Tirole 1994; Choi 2001).

7.2.2 Biobank Decisions

Figure 7.3 summarizes the tasks a biobank must carry out and the decisions it must make.

7.2.2.1 Strategic Positioning

The first decision a biobank must make concerns its *strategic position* in terms of the scope (pathologies/organs included) of its collections and the quality of the samples it supplies. Whether a biobank specializes in a small number of collections or offers a wider range impacts both the amount of competition between biobanks and the biobank's expertise. Choosing to specialize in just one or two pathologies/organs gives a biobank a unique position in the scientific community and ensures it gains a high level of visibility and recognition. Focusing on a single area also enables a biobank to reduce its operating costs and develop a greater depth of expertise. Specialized biobanks can also enhance their collections by buying samples from other institutions (clinics, tumor banks, other biological resource centers). What is more, specialization places a biobank in a niche, thereby giving it a certain amount of monopoly power. Conversely, a biobank may decide to maintain a wide range of collections and thereby give itself the opportunity to contribute to a larger number of projects. Although such generalist biobanks can provide broad expertise, this expertise will be less detailed than that provided by specialized biobanks. In addition, the

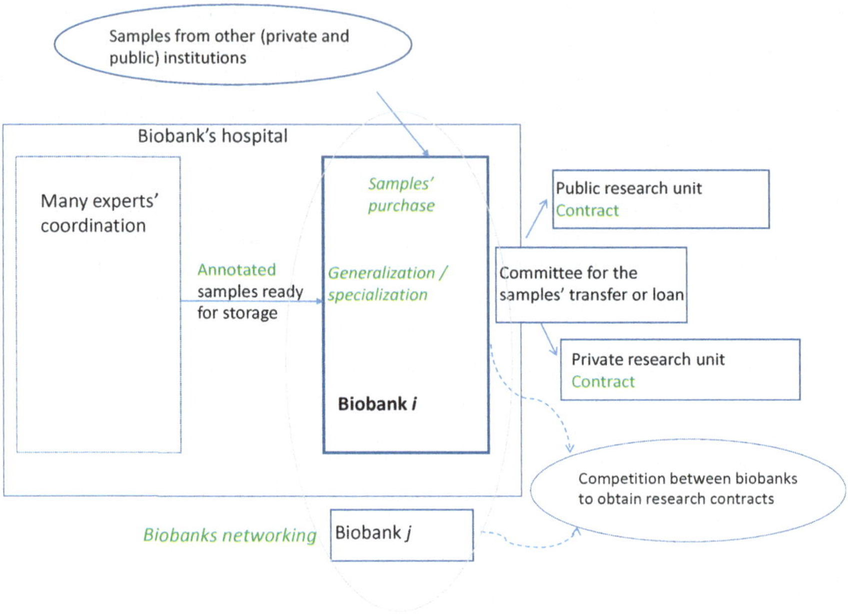

Fig. 7.3 Biobanks' different decisions

need to employ experts in many different fields increases the operating costs of generalist biobanks.

The second strategic decision concerns samples' quality. To optimize their future use, samples must be annotated very carefully when they are taken. A biobank can add supplementary information to these annotations when samples are transferred to a research unit. Providing high-quality samples allows a biobank to vertically differentiate itself from its competitors.

Biobanks' strategic choices can result in three types of economic landscape, as outlined in Fig. 7.4. The landscape in which a biobank operates will impact the way it valorizes its collections.

First, the landscape may be composed entirely of *generalist biobanks*, which compete to attract projects. One way of reducing competition within this type of landscape is through vertical differentiation based on sample quality. Conversely, different biobanks align themselves with different projects according to the quality of the samples required. This type of differentiation gives all biobanks some market power (Choi and Shin 1992; Motta 1993). Another possible configuration is for the landscape to be composed entirely of *specialized biobanks*. Competition within such a landscape is limited because specialization offers local monopolies to individual biobanks. Consequently, sample quality is less of a decisive factor when choosing a partner biobank because focusing on just one or two collections gives each biobank a different market position. Finally, *generalist and specialized biobanks* can coexist. This results in increased competition to the detriment of generalist biobanks, which

<table>
<tr><th></th><th>generalist biobank</th><th colspan="2">specialized biobank</th></tr>
<tr><td>generalist biobank</td><td>- strong competition
- low revenues
- outcome : vertical differentiation with several samples' qualities</td><td colspan="2">- weak competition but on the common collection
- outcome : - vertical differentiation on the common collection
- high quality for the specialized biobank</td></tr>
<tr><td rowspan="2">specialized biobank</td><td rowspan="2">- weak competition but on the common collection
- outcome : - vertical differentiation on the common collection
- high quality for the specialized biobank</td><td>same specialization</td><td>different specializations</td></tr>
<tr><td>- strong competition
- low revenues
- outcome : vertical differentiation with several samples' qualities</td><td>- no competition
- high revenues
- vertical differentiation is less important</td></tr>
</table>

Fig. 7.4 Biobanks' strategic positioning

may react by deciding to produce lower quality samples to reduce their costs. Conversely, specialized biobanks can profit from the situation by using their expertise to produce high-quality samples and thereby differentiate themselves from their generalist competitors.

7.2.2.2 Conditions of Transfer

Transfers cannot take place if a biobank refuses to provide samples. This decision is made by an access committee composed of scientists who assess applications from research units. The access committee may also provide guidance on whether the biobank should contribute expertise beyond the provision of samples. Biobank's expertise is a crucial variable, as it is easier for a biobank to assess a project's merits if it has specialized expertise in the field in question. Whether a research project is publicly or privately funded may also influence biobank's decisions. For example, it can be easier for public biobanks to work with research teams whose goals are to produce scientific knowledge without seeking any direct financial compensation for their research.

Once a biobank has approved a project, the biobank and research team must agree on the terms and conditions of the sample transfer. Under the resulting contract, samples may be transferred in return for a lump-sum payment, a royalty, or a tariff per sample. Contracts must be based on verifiable variables. Economically, it is important to determine the type of contract and the contract terms that will do the most to ensure research projects are successful. Whether the biobank will share in the proceeds from any resulting innovation is another question that must be addressed. The answer to this question will often depend on each party's bargaining power, that is, the alternatives they have (another project or partner) if negotiations fail.

A research project's nature and objectives, and especially the extent to which these aspects can be specified in a contract, are a fundamental element in the economic analysis of contracts between biobanks and their partners. It is very difficult to define each party's role in ensuring a project is successful. From an economic perspective, it is important to understand how the terms of the contract between a biobank and a research unit affect a research project's value.

Biobanks can use the contract *type* to convince potential partners of their quality and thereby to reduce the problem of adverse selection. For example, a biobank can signal its quality by proposing a contract in which it receives a lump sum payment at the time of signing plus royalties and a lump sum payment if the research project succeeds (Macho-Stadler and Perez-Castrillo 1991).

The success of a research project depends on the characteristics of the partners and on the efforts made by each party during the execution of the contract. From this point of view, the *terms* of the contract are important because they can influence the amount of effort each partner contributes to the relationship and, therefore, the extent of the moral hazard (Aghion and Tirole 1994; Choi 2001; Dechenaux et al. 2011).

One particularity of research projects is that they rely on the provision of a tangible good: the sample. Thus, most contracts specify a tariff at which samples will be made available. Biobanks must set this tariff in relation to the marginal cost of production. In practice, biobanks often engage in price discrimination, offering lower tariffs to public research units than to private research teams. According to economic theory, such price discrimination is inefficient (Tirole 1988). In this complex relationship, tariffs also have an impact on the involvement of actors, and this involvement may justify such discrimination.

If the contract can specify all the obligations of all the parties, it is possible to obtain the optimum level of effort via a contract specifying a lump-sum payment from the research team to the biobank (the result of the relationship is independent of the size of the lump sum). If the contract cannot specify all the parties' obligations, a fixed payment is no longer optimal and the biobank must be compensated for the effort it provides.

7.2.2.3 Networking

Although networking initially entails substantial standardization and coordination costs for a biobank, it can greatly increase its visibility, improve its access to samples, and, through the standardization of production processes, reduce its operating costs. In addition, networking reduces competition, thereby giving biobanks greater bargaining power in their negotiations with research units. On the other hand, the extra information asymmetries that can result from networking can lead to additional costs for biobanks. Hence, networking will be successful only if all the partners show a sufficient level of involvement. Inevitably, a biobank's involvement will be affected by the fact that it is no longer alone in contributing to a project's success. This dilution of individual responsibility can lead to freeriding problems should one member of a network try to profit from the work of its partners by leaving them to supply the necessary involvement.

In the case of voluntary networking, it is important to understand what is at stake for a biobank. Expertise is a key issue: It is more difficult for a specialized biobank to justify participating in a network because it would have to share its samples, as well as its expertise, with less specialized biobanks and therefore accept that it will lose a certain amount of control over its collection. Biobanks with greater expertise will also be more vulnerable to freeriding behaviors. Nevertheless, joining a network may still be attractive for specialized biobanks if it gives them access to research projects they would not be able to manage on their own. Hence, biobanks will join networks only if they can achieve added value by sharing samples or collections (Casamatta and Haritchabalet 2007).

The positive effects of networking are greater for generalist biobanks. The resulting increased visibility allows generalist biobanks to participate in more, and more ambitious, projects. Moreover, collaborating with specialized biobanks enables generalist biobanks to increase their expertise, which, in the long term, has a positive effect on the efficacy of research projects (Tykvová 2007).

7.3 Conclusion

An analysis of the interactions between strategic positioning, sample-transfer conditions, and information asymmetries shows that expertise is a key factor in both the economic valorization of biobanks and the success of research projects. In fact, expertise determines both a biobank's position on the quality scale and the amount of competition between biobanks.

References

Aghion, P., & Tirole, J. (1994). The management of innovation. *The Quarterly Journal of Economics, 109*(4), 1185–1209.

Casamatta, C., & Haritchabalet, C. (2007). Experience, screening and syndication in venture capital investments. *Journal of Financial Intermediation, 16*(3), 368–398.

Choi, C. J., & Shin, H. S. (1992). A comment on a model of vertical product differentiation. *The Journal of Industrial Economics, 40*, 229–231.

Choi, J. P. (2001). Technology transfer with moral hazard. *International Journal of Industrial Organization, 19*(1), 249–266.

Dechenaux, E., Thursby, J., & Thursby, M. (2011). Inventor moral hazard in university licensing: The role of contracts. *Research Policy, 40*(1), 94–104.

Macho-Stadler, I., & Perez-Castrillo, D. (1991). License contracts and information asymmetry. *Annals of Economics and Statistics*, 189–208.

Motta, M. (1993). Endogenous quality choice: Price vs. quantity competition. *The Journal of Industrial Economics, 41*, 113–131.

Tirole, J. (1988). *The theory of industrial organization*. Cambridge, MA: MIT Press.
Tirole, J. (1999). Incomplete contracts: Where do we stand? *Econometrica, 67*(4), 741–781.
Tykvová, T. (2007). Who chooses whom? Syndication, skills and reputation. *Review of Financial Economics, 16*(1), 5–28.

Chapter 8
Socializing Tumors: From the Conservation of Tumors in Banks to Their Ontological Variations

Fabien Milanovic

This paper questions the conservation of living organisms in banks and their uses from a unique perspective in view of the one that is often borrowed and destined to contribute to improve the practices in question. Since the operation of biobanks raises the defining issues of technical and ethical procedure choices, such a practical perspective seems obviously justified (Chassang et al. 2015; Beier et al. 2011; Noiville and Bellivier 2009; Gottweis and Petersen 2008). Biobanks are thus considered as developing infrastructures calling for a joint political, ethical and legal regulation. They are a focus for action. But how do matters stand when biobanks are a sociological focus of research, released from performative issues, when "the social" is not seen as the equivalent of "society" which causes problems or generates constraints in relation to the access and the use of parts of the human body? A theoretical approach to the social as an attachment process between human and non-human entities (a well-proven perspective in the field of *Science Studies*, cf. Latour 2005) raises the question of biobanks being considered as a body of living organisms socialization, meaning a place where human and non-human entities are associated, where the human body is broken up, fragmented (Rose 2007) into various parts destined to circulate within different environments (for example clinical, academical, industrial…) and designed to achieve a variety of objectives (for example medical, scientific, economic…).[1]

[1]For a critical review of the literature see Lafontaine (2014).

F. Milanovic (✉)
Sup'Biotech Paris, Paris, France
e-mail: fabien.milanovic@supbiotech.fr

X. Bioy (ed.), *Public Regulation of Tumor Banks*,
https://doi.org/10.1007/978-3-319-90563-1_8

8.1 The Socialization of Living Organisms as a Focus of Research?

Questioning the conservation of living organisms in banks implies a growing acuity since, in the last few decades, there has been a proliferation of various ways of collecting human body parts: blood, plasma, DNA, RNA, protein, organ, tumor, egg and sperm banks, which might be of cell line, and others (for a comprehensive overview, see Milanovic et al. 2017). However, we know very little about what happens between the removal and the use of the resulting elements in various activities, be they medical (diagnosis, therapy, prognosis) or the object of biomedical research to identify new biomarkers or conduct clinical or longitudinal—cohort—studies.

Admittedly, the study of relations between humans and non-humans has been the object of a vast number of works (Houdart and Thiery 2011; Barbier and Trepos 2007; Callon and Law 1997), focusing on objects or things, but the "non-human" category is diverse enough for a more thorough approach which might raise new questions: by making a distinction among non-humans—between living and non-living entities—questions may focus on the sequencing of different types of living entities (Milanovic 2015); it is also possible to differentiate, within the human species, between human bodies and the elements sampled from them to be used.

Within this framework, I wish to explore the socialization of tumor being through the current specificity in biobanking activities. Bank conservation implies the extraction of bodily elements which belong to a body, but what kind of link still exists between those elements and their body of origin? How do samples become a resource to act upon?

8.2 A Sociological Fieldwork on Tumor Biological Resources

To answer the questions raised, I shall draw on the sociological fieldwork I conducted in France on cancer tumors (Milanovic 2008). I shall endeavor to highlight the fact that, as any resource, human biological resources are *fabricated* (which will entail some consequences), and not "natural" at all (if they were, they would already be there, ready for collection and further use). As the etymology of the word resource[2] suggests, a process is necessary to allow re-use of elements extracted from the human body. To remember that fact one needs to emphasize the "artifactual" nature of resources. Such an approach to biological resources enables us to understand their socializing transformation: how are they built and

[2]Re-source, *re-surgerer* in Latin; to appear again.

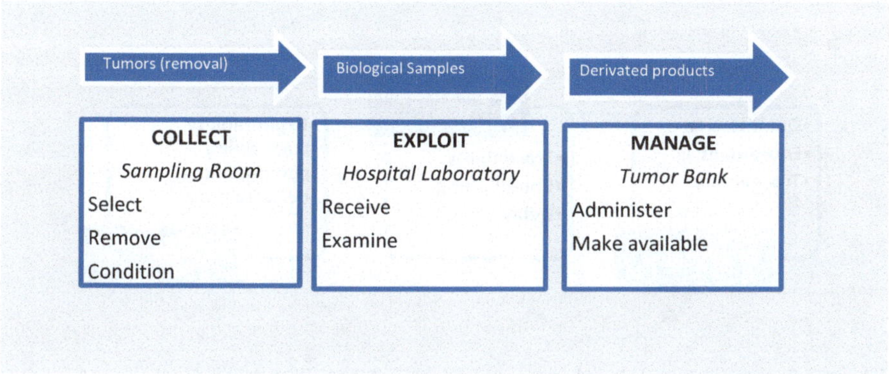

Scheme 8.1 Sociotechnical chain in tumors biobanking

turned into a state that enables humans to use them as they become part of the collective?

Socializing living, out-of-body, fragmented organisms, is not just about complying with standards—technical or legal/ethical—to authorize their circulation and use in different environments, it is also about shaping them to that end. Living organisms are not a monolithic and immutable entity, the circulation of which should be established. Socializing them is about shaping them, turning them into hybrid entities, a process which leads to a biological, technical and social entwining (Helmreich 2015; Vermeulen et al. 2012). Drawing on the empirical data collected during the fieldwork, I shall focus on removed tumors and follow them from one phase to another, from their removal from the human body to them being turned into biological resources ready to be used in the medical or research field. That course implies the creation and implementation of a real **sociotechnical chain** enabling biobanking in hospitals which takes place in three successive "scenes" (sampling room, hospital laboratory, tumor bank) and entails a great number of actions as is indicated in Scheme 8.1.

Through the description of the phases which enable biobanking (namely the collection, the exploitation and the management), I shall demonstrate that tumors undergo a series of tests which makes them suitable for further use. I shall argue that biobanking implies shaping actions which generate an **ontological plurality** of samples—they exist in a variety of ways. Indeed, from the tumor itself to its samples and by-products, bodily elements are increasingly mediated, their link to their body of origin is less direct, which opens up a space for their various existences—ontologies—where the link to the body—maintained or severed—makes the status of the sample entity vary legally from 'being' to a 'thing', which multiplies its potential uses.

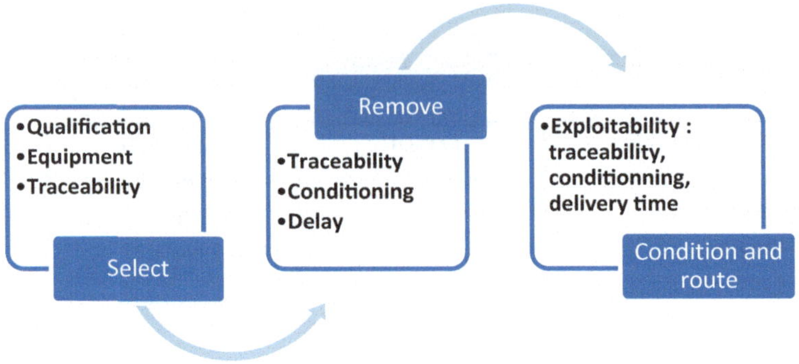

Scheme 8.2 Sociotechnical chain in tumors biobanking at the collection stage in the sampling room

8.3 The Sociotechnical Chain of Biobanking

This part will deal with the procedure of tumor biobanking via the sociotechnical chain in use. It comprises three stages (which are linked): **collecting** tumors, **exploiting** them and **managing** them, each stage comprising numerous actions. Each stage will be depicted in the following sections below. Those three stages are parts of the creating process of biological resources. The idea that I put forward is that the transformation of bodily elements into biological resources takes place through mediations meant to qualify them, equip them and organize their traceability to make them suitable for future use.

8.3.1 Scene 1: Collection

As seen in Scheme 8.2, the tumor biobanking sociotechnical chain comprises four steps involving the collection of the sample in the hospital theatre: selecting tumors, removing them, conditioning them and delivering them. Let us examine each step.

8.3.1.1 Selecting Tumors

Selecting tumors depends on political and institutional procedures which link governmental measures—Cancer Plan in France—to the Health and Research government departments, and to public and private hospitals which are part of the national health service—such as Cancer Centers—where biobanks[3] are located. During the

[3]There are about 60 of them in France. For more information, cf. *Les tumorothèques hospitalières. Recommandations à l'usage des cliniciens et des chercheurs*, Collection "Recommandations professionnelles" published by Institut National du Cancer (Inca), Boulogne-Billancourt, 2006.

collection phase in hospitals, tumors are selected both for medical and research purposes, with the former being prioritized over the latter (research is only possible if some samples are left after medical tests). Within a precise legal and ethical framework, tumor collection involves organization and traceability: the quality of biological samples, comprehensive patients' data (clinical, biological, anatomical and cytopathological), and the assigning and coordination of roles in the collection process within the biobanking sociotechnical chain. From the first set of actions linked to collection, it is important for biobankers to have samples which are **qualified** for purpose—for medical and research purposes, in compliance with the law; which are **equipped**—patient's file, forms, importance of the site and of the hospital equipment which enables cryopreservation; and well **tracked**—systematically and dynamically, when the patient is part of a cohort studied over many years.

8.3.1.2 Collecting

Nucleic acids, which are parts of tumors, should be of good quality as they are essential to pursue the necessary molecular analysis recommended by the french INCA (Institut National du Cancer) prior to diagnosis. The exploitability of the sample will depend on several factors which should, therefore, be considered during the collection phase.

The first factor is the available information giving samples the relevant qualities for future uses and users; it is also paramount for the rest of the course. To that effect, traceability should be required. It is even more necessary, as not only are bodily elements of the patient collected, but also personal information—clinical, biological, anatomopathological—and details pertaining to hospital care—notably the time of collection. The quality of samples and, by extension, of the collections of tumors, is directly dependent on it.

Organizationally, attention is drawn to "cooperative work" as it refers to the additional work that must be done so that tumor biobanking ends up being more than just a "chaotic assembly of scattered fragments of work." (Strauss et al. 1992, p. 191). The management and formatting of the trajectory of the samples in the hospital are impacted by this work. There must be standardized operational procedures to limit hazards, articulating the various sequences of work involved and, more generally, to align resources and entities involved in the collection work. In this respect, the first person handling the sample, the surgeon, must associate the act of logging written information, including the time, with the technical act of extracting the tumor, in a standardized document (the "test request" in the case of University hospitals). The surgeon should therefore be "aware" of that. The need for these practices to meet standards is made difficult by the fact that the hospital site is vast in terms of size and services, making communication activities that necessary for such awareness more complex.

These practices should be highlighted as they are at the heart of concerns of biobankers in terms of traceability: the necessary information should come with tumor samples to allow their characterization with properties that are adjusted to their future uses—medical and scientific. This will give them value.

8.3.1.3 Conditioning

The second important factor at this stage of exploitability is related to the means used to protect the tumor, specifically conditioning. Nucleic acids of tumor tissue are likely to deteriorate within a very short time from the moment the tumor is extracted. The nature of the conditioning used is therefore crucial. Either freezing is immediate on the site of collection (it is the "gold standard"); or else it is postponed and the sample is provisionally maintained in a stabilizing solution such as 'RNAlater' as it is waiting to be delivered to the pathology laboratory. However, the solution is likely to affect the quality of the nucleic acids. There is another option: the sample is conditioned as "fresh."

The third determining factor—in terms of exploitability of the sample—is the amount of time passed prior to its delivery to the lab for solid and tissue samples (not for blood tumor samples, which can be used for up to 24 h). For them, the time between collecting and freezing should be very short. This directs us to the last phase of collection: the delivery.

8.3.1.4 Delivering

Transportation points to the spatial dimension inherent in any collection chain: how are the tasks involved in biobanking distributed in space? Does the territory of collection go beyond the boundaries of the hospital site, or does it stand within it? The terms of distribution of the chain depend, of course, on the area of collection of the targeted tumors, and, therefore, on the collection policies in place. It is worth noting that the area covered by the site and its importance complicate the modalities of coordination between the activities of collection and delivery to the laboratory. The implantation on a single site in the biobanking chain facilitates any information retrieval that could be undertaken later. By having, within a small space, entities involved in the chain, biobankers are more able to control the route of the tumors and to track actions that take place. But what about other cases where samples are collected as part of a multi-center Hospital Clinical Research Program (HCRP), the recruitment area of which is on a national-scale?

Freezing should be completed in optimal conditions where molecular biology analysis must be undertaken at the level of nucleic acids from sampling, whether for medical or scientific purposes; but this is not always the case. The exploitation of tumor samples, which is the next phase of the biobanking chain, directly depends on what has been done upstream, i.e. during the first phase of collection. In this respect, one of the disadvantages of multi-center research, even carried out

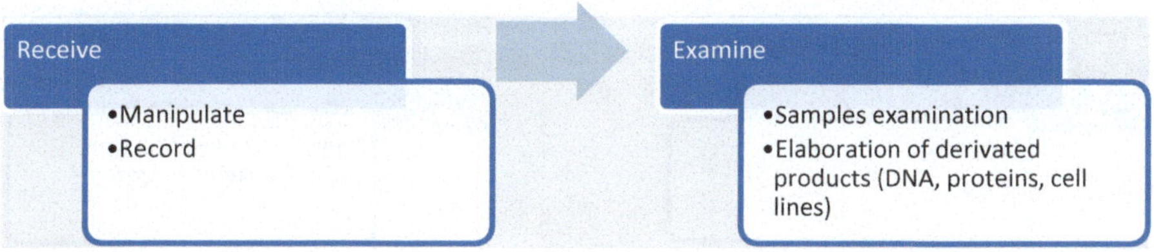

Scheme 8.3 Sociotechnical chain in tumors biobanking at the exploitation stage in the hospital laboratory

under a HCRP—with the protocol as a prescriptive tool, is a lack of "grasp"[4] on the operations of collecting and conditioning tumors that the chain enables the biobanker to undertake.

In the end, the conditions of collection, on which depend the quality of nucleic acids, introduces a hierarchy of biological samples, through the different means of control according to the quality approach people are willing to use. The inherent traceability in this approach provides a lot of information available in computability operations—e.g. type of conditioning, delay between sampling and conservation, conditions of delivery with a standardized medium for the information such as "medical care form", "computerized patient record", "test request"—carried out to assess and assign a differential value to biological samples and the collections they are part of. In that sense, the spatial distribution of the tasks of biobanking is fundamental and should be carefully embedded into its organization and its reports. The deployment of the collection of tumors depends on organizational work, on the spaces where these activities unfold, and on the characteristics of sites of collection, especially in terms of equipment and infrastructure.

8.3.2 Scene 2: The Exploitation in the Hospital Laboratory

As shown in Scheme 8.3, the phase when pathology laboratories receive tumors comprises many actions: manipulating and splitting samples, classifying and storing tumor samples, but also entering, searching, categorizing, calculating and recording data on associated tracks. It is therefore a process involving two aspects: firstly, actions that are intended to form the collection. In this respect, manipulation consists in fragmenting the sample and assigning various purposes to the resulting aliquots, the purposes being medical (fixed and frozen samples) and scientific (frozen samples using several techniques of cryopreservation, and to a lesser extent fixed samples). Then, other actions will concern related tracks. On this second aspect, data and annotations of the sample are digitized in a computer database that is managed by a dedicated software.

[4]For a theoretical perspective on the notion of grasp, see the work of Bessy and Chateauraynaud (1995, reissued in 2015).

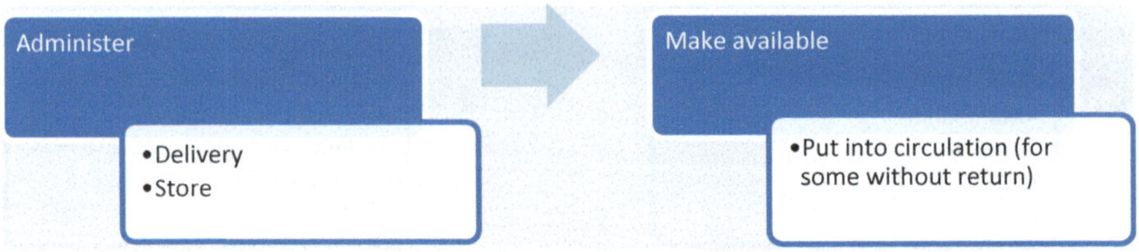

Scheme 8.4 Sociotechnical chain in tumors biobanking at the management stage in the tumor bank

During biobanking, the handling and recording of samples and associated information mean there are variations on the formatting of the resulting samples and the nature of annotations—clinical, biological, anatomo-cytopathologic, and dynamic-, which again contribute to the quality and the usability of tissue samples. As the referent technician of the cell bank of St. Louis puts it: "The sample has a value, its quality relies on the annotations that are attributed to it, but also on the freezing process. You can freeze a sample appropriately, but if it is not annotated, it has no value. So, the annotations are really very, very important".

This process of formatting, not during the collection, but in relation to the exploitation phase (here in the reception phase), through the assignment of properties, participates in the formation of the value of the tumor samples. The objectified quality of the samples in biobanks, and their computerized records, offer ways for collectors and users to assess and assign a differential value to these biological samples (e.g. cell count and viability as "quality control"), and to use them. Once the sampling is formatted, the resulting "biological samples" and the information related to them can be used in the prescribed tests, which will in turn give them other properties.

In this second stage of tumor banking, the work organization in terms of what handling, recording and examining involve will occur in the process of formatting the samples and their tracks. Through this formalization and testing, tumor samples are qualified in terms of use and in terms of the tests they have undergone, equipped—kept in a stable state—and tracked—controlled and monitored. One can note that this formalization gives properties related to samples which increase their potential subsequent connections for further use. But for now, they leave the pathology lab and go to the tumor bank where they are managed.

8.3.3 Scene 3: Managing Tumor Entities

Scheme 8.4 shows what managing tumor entities entails. It is a complex activity with many ins and outs, but put simply, to manage is to **make available**, both internally (i.e. at the level of the people involved in the biobank or the pathology department), and externally (requests for samples from outside the biobank or the hospital site). Referring to works published for more details (Milanovic et al. 2007;

Pontille et al. 2007), let's just say that these requests from different sources are dealt with under differentiated exchange plans: according to the nature of the relationship between the protagonists, the geographical location of the applicant (closest to furthest, from internal to extra-European applications) and his status (public-private), the release of biological entities can take place as a donation, as a partnership or as a service or goods delivery.

This is the first aspect in terms of actions of what "making available" means. There is a second aspect, related to the fact that, although essential to medical and cancer research, tumor biological resources are available in small quantities (biopsies, micro-samples). Like micronutrients present in very low amounts are essential to metabolism, tumors contain biological oligo-resources. And one suspects the finite quality of the available quantity exacerbates the management methods of their uses.

However, another part of these resources is replicable: indeed, derivatives can be made from nucleic acids using techniques of molecular biology (DNA, protein) or immunophenotyping (cell lines and immortalization). These products (DNA, RNA, proteins, cell lines) are then ready to be used in various actions, be they medical (thorough testing) research-based or related to biobanks both internally and externally. In this sense, they are "made available".

To summarize, once collected in a hospital department and brought to the pathology laboratory, tumors are split and, consequently, subdivided in several samples. The organization of the underlying work consisting in manipulating, saving and examining is about formatting these samples and their tracks. One of them is reserved for medical use; it is "the patient's" as the biobank technician said. Others can be used for research, as long as they have been qualified for this. The fragmentation of a sample means it is possible to assign differentiated purposes to a tumor. Such actions affect the form of the entities and the state of their relationship, the form allowing different relationships according to the properties of the sample. Samples, conditioned in a certain way and preserved by cold or fixed in paraffin, then leave the laboratory and enter the biobank where they are managed. At this last stage, in the biobank, some elements of the samples are reportable and reproducible at will. Derivatives can be made from nucleic acids using techniques of molecular biology (DNA, protein) or immunophenotyping (cell lines and immortalization).

8.4 Making Biological Resources: An Ontological Plurality

Let us consider the movement of mediation and its implications: formatted samples from the collected samples take a first series of medical tests, possibly followed by secondary tests. They are then likely to be used for research purposes, which can be designed as a detour back to the patients—it is the case for clinical and translational research or as an end in itself if it is "fundamental" research. From the first exploitations of the tumor sample to the last, operations move away, gradually, from personal or direct interest to the patient. Conversely, the relevance for the

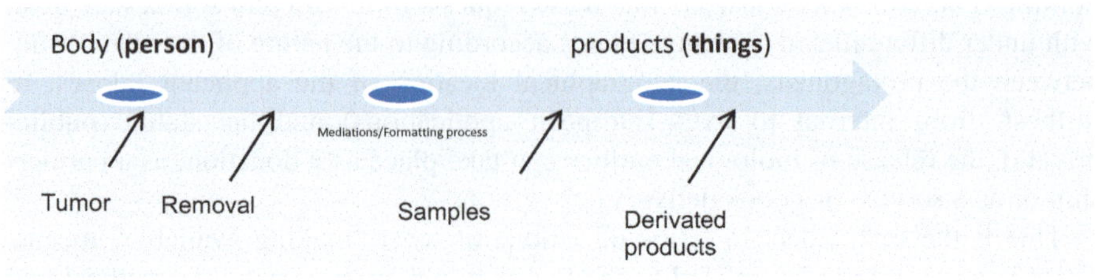

Scheme 8.5 Formatting of tumors in biobanking

patient in terms of the use of their sample gradually decreases. The sample also follows this detachment route: from the tumor in the patient's body (first formatting) to biological samples, and from biological samples to derivatives (second formatting), the elements of the human body are increasingly mediated until they are of a different nature at the end (as the word suggests, derivatives are not from the patient's body, but are derived from it). Scheme 8.5 shows this formatting of tumors in biobanking.

Reviewing the various steps that punctuate biobanking highlights how biological resources, before being points of departure for biomedical actions, are points of arrival in the sense that they result from activities. These activities, articulated within a sociotechnical chain, allow transformation of the tumor, from one form to another: from the tumor to the sample, which is then split into further samples and subsequently turned into derivatives. When these transformations occur, tumor entities are given associative or connectionist properties. This set of transformations allows the tumor source to become a biological re-source, as it has been formatted as such. Therefore, it is through the transformation of native tumors into tumor entities within a sociotechnical chain of mediations that samples are transformed into biological resources, where, in other words, biological resources are made.

Understanding the social as a movement which consists in re-associating or reassembling, it seems that such work is to socialize cancer tumors since this work of formatting is a project which aims to associate them with other entities, whether *human* (patients, clinicians, researchers, industrialists and SMEs), *instrumental* (when tested with optical or electronic microscope, ultramicrotome and cryotome,[5] laser microdissection), *technical* (extraction of molecules, dosages, looking for markers, immunological studies), *institutional* (cf. public hospitals, Inca, DHOS, DRC, government departments and other academic institutions that administer or fund them) or in related *infrastructures* that host them in highly equipped

[5]The ultra-microtome is a device that performs cuts of 80 nm thick on samples included in blocks of very hard resin. The elements of the examined cell can then be magnified up to 100,000 times under a beam of electrons. As for the cryotome, it enables cutting at low temperatures (−25 °C) a sample, previously frozen, into thin sections (a few micrometers), which will be colored after having been placed on a transparent slide.

environments (biobanks, centers of biological resources, biological investigation centers, technology platforms, cancer centers...).

Biological resources thus result from the formalization of tumors, which means turning these into a stabilized and mobile form, with tailored properties for their future recipients and uses. The resulting tumor elements (samples, subsequent smaller samples, aliquots, derivatives, cell lines, oligo-resources, collections) are assembled, considering the different social worlds involved (patients, clinicians, researchers, hospital, government institutions) and those likely to invest in them (industries and companies, public authorities, associations, etc.). In this sense, their properties link tumors to individuals, institutions and heterogeneous objects: the layout of the properties of tumor entities—variously presented—gives them a connectivity that allows a multiplicity of *forms of existence*. One can therefore note an ontological plurality which can be ordered according to the link to the patient each form entails: we have seen that the mediations introduced during the course of formatting move the tumor further from the patient until their link is severed, thereby changing the tumor entity from being legally part of a person (maintained link) to a 'thing' (severed link).

Cancer tumors are not stable realities, defined once and for all by fixed and essential characteristics: they are always evolving, subjected to formatting sociotechnical processes that make them move from the status of living entities (linked to the bodies of the people they come from) to 'things' or 'bio-objects' (Webster 2012) partially reproducible (derivatives) or available in limited quantity (oligo-resource), as the case may be. A tumor can be broken into multiple ontological entities with differentiated capabilities of action. The legal framework, discussed in this text and more generally in this book, is part of the process: the link maintained between the patient and his tumor gives him rights over it—a right of withdrawal—while a severed link paves the way to a legacy, to potential transferability. In short, the actions of linking and severing a link are assigned to a legal regime—that which governs people-, or another—that which governs things-, some calling for a third—that of 'human things'; those regimes are related to other practices (commerciality, appropriation, etc.).

The complexity surrounding banking and the socialization of cancer tumors justifies the need to engage in reflection without considering practical issues, at least initially. That reflection should be multidisciplinary to define, with due regard to empiricism, the terms of existence of tumors within our world.

References

Barbier, R., & Trepos, J.-Y. (2007). Humains et non humains: un bilan d'étape de la sociologie des collectifs. *Revue d'Anthropologie des Connaissances, 1*(1), 35–58.

Beier, K., Schnorrer, S., Hoppe, N., & Lenk, C. (Eds.). (2011). *The ethical and legal regulation of human tissue and biobank research in Europe*. Göttingen, Germany: Universitätsverlag Göttingen.

Bessy, C., & Chateauraynaud, F. (2015). *Experts et faussaires. Pour une sociologie de la perception*. Paris: Petra. (1st issue in 1995).

Callon, M., & Law, J. (1997). L'irruption des non-humains dans les sciences humaines: quelques leçons tirées de la sociologie des sciences et des techniques. In *Les limites de la rationalité (tome 2), Colloquium of Cerisy (under the supervision of Bénédicte Raynaud)* (pp. 199–218). Paris: La Découverte.

Chassang, G., Cambon-Thomsen, A., & Rial-Sebbag, E. (2015). *Ethique et réglementation des biobanques de recherche*. Paris: Biobanques Inserm US013.

Gottweis, H., & Petersen, A. (2008). *Biobanks. Governance in comparative perspective*. London: Routledge.

Helmreich, S. (2015). *Sounding the limits of life: essays in the anthropology of biology and beyond*. Princeton, NJ: Princeton University Press.

Houdart, S., & Thiéry, O. (Eds.). (2011). *Humains, non humains - Comment repeupler les sciences sociales*. Paris: La Découverte.

Lafontaine, C. (2014). *Le corps-marché. La marchandisation de la vie humaine à l'ère de la bioéconomie*. Paris: Seuil.

Latour, B. (2005). *Reassembling the social*. Oxford, England: Oxford University Press.

Milanovic, F. (2008) *La socialisation des tumeurs cancéreuses. Une enquête sociologique des tumorothèques françaises* (research report). Toulouse, France: Inserm U558, funded by INCa.

Milanovic, F. (2015), Utilitarisme et rapports à la nature: la fin d'une démesure ? Sur les modes d'existence et l'agentivité du vivant. In *"Démesure", 2nd international Congress of AFEA, 29 June–2 July 2015, Université Toulouse-Jean Jaurès, Campus du Mirail, Toulouse*. Online: https://demesure.sciencesconf.org/156876/document (consulted on 6 July 2017).

Milanovic, F., Pontille, D., & Cambon-Thomsen, A. (2007). Biobanking and data sharing: A plurality of exchange regimes. *Genomics, Society and Policy, 3*(1), 17–30. Downloadable article: http://www.gspjournal.com/ (consulted on 6 July 2017).

Milanovic, F., Merleau-Ponty, N., & Pitrou, P. (2017). Biobanques: quelles reconfigurations pour le vivant? Approches interdisciplinaires. *Natures Sciences Sociétés, 25*(3), 268–275.

Noiville, C., & Bellivier, F. (2009). *Les biobanques*. Paris: PUF.

Pontille, D., Milanovic, F., & Rial-Sebbag, E. (2007). Collectionner le vivant: régulation, marché, valeur. *Revue d'économie industrielle, 120*, 195–212.

Rose, N. (2007). *The politics of life itself*. Princeton, NJ: Princeton University Press.

Strauss, A., Fagerhaugh, S., Suczek, B., & Wiener, C. (1992). Le travail d'articulation. In A. Strauss (Ed.), *La trame de la négociation* (pp. 191–244). Paris: L'Harmattan.

Vermeulen, N., Tamminen, S., & Webster, A. (Eds.). (2012). *Bio-objects: life in the 21st century*. Farnham, England: Ashgate.

Webster, A. (2012). Bio-objects: exploring the boundaries of life. In N. Vermeulen, S. Tamminen, & A. Webster (Eds.), *BioObjects: life in the 21st century* (pp. 1–10). Farnham, England: Ashgate.

Part II
The Issues Related to Tumor Collections Valorization

Chapter 9
The Valorization of Tumor Collections: Reconciling the Interests of the Different Actors

Emmanuelle Rial-Sebbag

Lawyers have taken an interest in biobanks ever since they first emerged onto the clinical and research landscape (Rial-Sebbag 2009; Bellivier and Noiville 2009). The main concern at this time was to protect donors' and research subjects' rights with respect to the creation of collections and the use of biological samples and associated personal data. Consequently, legal and ethical issues were closely related to individuals' rights (consent, information) and safety. However, recent years have seen substantial changes in the way biobanks operate, as the samples they house are now routinely used for research as well as for patient care. Because samples are used for multiple purposes and to achieve a variety of objectives, which may be very different from the original reasons for harvesting them, donors may be at several removes from decisions taken about their samples. Nevertheless, guaranteeing donors' fundamental rights is just as important as making best use of every sample. Consequently, the legal focus is now on transfers of existing samples, institutional partnerships, and feedback to donors or to professionals working on the collections.

A key issue in the interactions between these actors is the transfer and exchange of biological samples and associated data. Although the notion of "valorization" is usually viewed in financial terms (this is also the first definition of the term given in most dictionaries), the valorization of collections of biological samples must be considered more globally and embrace other values, including social and scientific values. Of course, financial value and other forms of value are not mutually exclusive. In France, for example, the laws covering bioethics forbid the commercialization of the human body, but biological samples can be subject to financial transactions and be used in projects that may lead to financial benefits through tools such as patents. Thus, collections of biological samples are by no means static; they circulate and may be subject to contracts. The need for these movements to consider

E. Rial-Sebbag (✉)
UMR 1027 Inserm, University of Toulouse III Paul Sabatier, Toulouse, France
e-mail: emmanuelle.rial@univ-tlse3.fr

X. Bioy (ed.), *Public Regulation of Tumor Banks*,
https://doi.org/10.1007/978-3-319-90563-1_9

the interests of all the actors involved raises several potentially contentious issues over aspects such as ownership, control over samples and data, and rewards for donors (feedback, remuneration, etc.). Indeed, although patients' rights must be a central concern when valorizing collections (I), scientific and economic imperatives (II) are also of paramount importance. These two issues are often considered to be conflicting.

9.1 Patients' Rights: A Central Concern When Valorizing Collections

Attention must be paid to protecting patients' rights when setting-up and valorizing tumor banks. Most samples are taken for treatment purposes and then repurposed for research, a change of status (patient becoming research subject) that donors generally do not understand clearly (Rial-Sebbag 2006). In France, the legal framework relating to this area is fragmented because it draws on laws and regulations from different fields.[1] Nevertheless, all these laws and regulations aim to ensure donors are properly informed of the uses to which samples and data will be put so they can be considered autonomous persons, even if these prerogatives cannot be assimilated to a property right (Le Corre et al. 2016). Donor autonomy means that individuals have the right to give their consent to samples and data being taken, but they may also refuse to allow sampling. The legal rules linking source persons to users are based on two fundamental principles: donation for research and authorization for future research. Thus, in the case of tumor banks, these principles directly concern the medical practitioners and health professionals in charge of collecting samples, who are considered responsible for the collection (Sect. 9.1.1), and indirectly concern the heads of Biological Resource Centers (BRCs), who are responsible for storing and managing the samples and data that constitute the collection (Sect. 9.1.2). Medical practitioners are the only people in these two categories who have direct contact with patients.

9.1.1 Respecting Patient Autonomy in the Medical Relationship

Social solidarity is the main motivation for patients' decisions to donate biological samples to research. In fact, most tumor samples are collected as part of a treatment program that will benefit the patient (1), who may later be asked to allow residues

[1]Loi n° 2002-303 du 4 mars 2002 relative aux droits des malades et à la qualité du système de santé modifiée, Loi n° 2012-300 du 5 mars 2012 relative aux recherches impliquant la personne humaine modifiée par l'ordonnance n° 2016-800 du 16 juin 2016.

from these samples to be used for research purposes (2). Hence, research programs do not usually benefit the sample donors directly, but the knowledge a research program generates may help other patients suffering from the same condition.

1. Medical practitioners generally ask cancer patients to authorize the storage of biological samples for diagnostic purposes. Tissue resection and blood samples are used to establish or confirm a diagnosis, or to determine whether treatment needs to be adapted.[2] These investigations involve biological or genetic analyses; therefore, they must be conducted in accordance with the law on patients' rights. Because biological tests are prescribed as part of a medical consultation, the patient must be informed of the test's purpose and of any risks involved, and give oral consent[3] to the test being carried out. Medical practitioners must inform their patients if genetic analyses will be carried out on any samples taken, and provide information about issues such as the possibility of informing relatives of the index subject[4] and the obligation to obtain written consent.[5] However, when samples are taken for treatment purposes, it may already be envisaged to use parts of these samples for research purposes. These situations are governed by specific rules.
2. Using samples for research purposes is a subject medical practitioners may broach with their patients either when taking samples for treatment purposes or at a later date. Under French law, patients may refuse to allow sample residues to be repurposed for research[6] (opt-out procedure). However, until recently the law was stricter than this in the case of genetic analyses, which, under common law, could only be carried out with the patient's written consent. In 2012, new legislation covering biomedical research, known as the Jardé Act,[7] modified this rule so that genetic analyses can be carried out unless a patient refuses consent for his or her samples to be used in research.[8] The medical practitioner

[2]According to the so-called "Pharmacogenetics".

[3]Article L1111-2 Public Health Code: "Everyone has the right to be informed about his health. This information relates to the various investigations, treatments or preventive actions that are proposed, their usefulness, their possible urgency, their consequences, the frequent or serious risks that are normally foreseeable, as well as the other possible solutions and the foreseeable consequences in case of refusal (...)".

[4]Article L1131-1-2 Public Health Code.

[5]Article 16-10 Civil Code: "Examination of the genetic characteristics of a person can only be undertaken for medical or scientific research purposes.

The express consent of the person must be obtained in writing prior to the completion of the examination, after being duly informed of its nature and purpose. The consent mentions the purpose of the examination. It is revocable without form and at any time."

[6]Article L1211-2 al.2: "The use of elements and products of the human body for a medical or scientific purpose other than that for which they were collected or collected is possible, unless opposed by the person on whom the collection took place, duly informed beforehand of this other purpose."

[7]Loi n° 2012-300 du 5 mars 2012 *op. cit.*, entered into force in 2016.

[8]Opt-out system implemented by Article L1131-1-1 Public Health Code: "Notwithstanding Article 16-10 of the Civil Code and the first paragraph of Article L. 1131-1 of this Code, the examination of the genetic characteristics of a person for the purposes of scientific research may be carried out from

must describe the potential benefits (individual or collective) of any research carried out using a patient's samples and the conditions under which the samples will be stored and transferred. To help researchers aiming to set up tumor banks, the National Cancer Research Institute (INCA 2011) issued several recommendations concerning general procedures. However, these procedures were adopted prior to the entry into force of the Jardé Act and therefore do not include the new rules for genetics. The provisions included in both the Civil and Public Health Codes are aimed at ensuring trust between patients and medical practitioners by making it a medical practitioner's duty to clearly explain to patients' possible future uses of their samples. Valorization issues are generally poorly addressed at this stage, although the scope of initial "consent" (in this case, non-opposition) largely determines a sample's future uses. Indeed, if a patient gives consent for his or her samples to be used only for research within a restricted field (e.g., the patient's pathology), these samples cannot be used for research outside this field unless the patient is consulted again (from a logistical point of view, re-contacting patients can be difficult or even impossible). Hence, patients have a globally recognized right of control over the use of their samples, based on information provided when the samples are taken and the extent of the consent they give for using these samples in research. Here, valorization is purely intellectual and consists of the donor's contribution to producing scientific knowledge.

BRCs, given the nature of their missions and their contribution to research, are subject to very different requirements.

9.1.2 Biological Resource Centers: In Charge of the Quality and Safety of Samples and Associated Data

According to the OECD, Biological Resource Centers (BRC) are "an essential part of the infrastructure underpinning biotechnology. They consist of service providers and repositories of the living cells, genomes of organisms, and information relating to heredity and the functions of biological systems. BRCs contain collections of culturable organisms (e.g. micro-organisms, plant, animal and human cells), replicable parts of these (e.g. genomes, plasmids, viruses, cDNAs), viable but not yet culturable organisms cells and tissues, as well as databases containing molecular, physiological and structural information relevant to these collections, and related bioinformatics" (OECD 2001).

Their mission has been defined as "conserving and (making) available in accordance with the legislation in force, collections of biological resources, particularly for purposes of research, heritage conservation, education or economic

---

elements of the body of that person taken for other purposes when that person, duly informed of the research project, has not expressed his opposition."

development".[9] Hence, a BRC's missions are to store biological samples in accordance with individual rights and safety rules (Sect. 9.1.2.1) and to ensure the utility of their collections by making them available to the scientific community for use in research (Sect. 9.1.2.2).

9.1.2.1 The Need for BRCs to Meet Biobanking Standards

Before a BRC can fulfill its missions, it must establish all necessary safety and regulatory compliance guarantees. The challenge for biobanks is to ensure their activities meet the operational rules of sample storage while respecting individuals' rights. Although BRCs must keep duplicates of statements of consent/non-opposition given on-site, links between donors and BRCs are tenuous. The law requires BRCs to follow several administrative procedures[10] (obtaining authorizations from and filing declarations with the Ministry of Research and Committees for the Protection of Persons) and they are encouraged to meet French quality standards for BRCs.[11] In terms of their operations, BRCs aim to standardize their sample storage methodology to facilitate the transfer of resources either to research teams within the BCR's parent organization (usually a public institution, hospital, research institute, etc.) or to outside research units (academic or private structures).

9.1.2.2 Providing Biological Resources

BRCs can only provide samples to researchers if research teams know the collections exist and if the transfer of the samples is authorized. Most BRCs have a website containing a catalog that can be used to view resources directly or indirectly (using login procedures). Requests and access are subject to an application and authorization procedure that includes evaluating the scientific merit of the applicant's project. Before samples can be transferred, initial consent records must be checked to ensure the will of the donor(s) is respected. A BRC's value is defined in terms of the volume of its activities and its ability to contribute to research projects.

Patients, health professionals, and BRCs have convergent interests because all the rules governing the collection, storage, and transfer of samples are aimed at maximizing the use of high-quality biological resources to produce scientific knowledge. Patients, that is, the donors of biological materials, want the greatest possible use to be made of the samples they provide (not using the samples could be considered as acting contrary to the patient's will) to produce knowledge that could benefit them (even indirectly). This is also the case for medical practitioners and researchers, who

---

[9]AFNOR, NF S96-900 Septembre 2011, Qualité des centres de ressources biologiques (CRB)—Système de management d'un CRB et qualité des ressources biologiques.

[10]Articles L1243-3 and L1243-4 Public Health Code.

[11]AFNOR op. cit.

want to improve their knowledge to provide better care to patients. BRCs, which work within a highly standardized framework, also contribute to knowledge production. However, new ways of valorizing collections and the emergence of new technologies are calling into question the relevance and effectiveness of these fundamental principles.

9.2 Reconciling Individuals' Rights with Scientific and Economic Imperatives

As noted above, the collection of samples can be considered successful only if the samples and associated data can be used for multiple purposes. However, protecting patients' interests is a central aspect of the law in the case of samples that can identify the person from whom they were taken.[12] Therefore, France's regulatory system gives patients the right to control how their samples and data are used. Although samples are not truly subject to property rights (Le Corre et al. 2016), patients must be informed of how their samples will be used and be consulted again if any research use is likely to go beyond the scope of the initial consent. However, even if research is carried out within the scope of the initial consent, patients have the right to feedback about the project and a right of access to the results. These ethical and legal obligations are not as clearly stipulated in French law on the use of collections as they are in the law on human research and, more problematically, they are incompatible with the development of a "market" in human biological material. Thus, it is necessary to look at how the interests of the different stakeholders in this market can be reconciled (Sect. 9.2.1) and the changes this market will generate with respect to the non-commercialization and ownership of human biological material (Sect. 9.2.2).

9.2.1 Developing a "Market" in Human Biological Material and Associated Data

The main purpose of the principles underlying France's 1994 law[13] on respecting the human body[14] was to ensure that patients and research subjects are kept fully informed about their medical treatment, the results of any tests carried out, and the outcome of any research in which they are involved. One of the system's main aims was to ensure a balance between an individual's medical interests and the needs of

[12]We acknowledge here that a link between the donor and his identification is still existing.

[13]Lois du 29 juillet 1994, op. cit.

[14]These principles are: primacy and dignity of the human person, respect, inalienability and non-commercialization of the human body, Article 16-1 of the Civil Code.

knowledge production. The rules relating to research were, and still are, based on the principle of donation, whether a study's results will be of direct benefit to its participants. Hence, the main reward for a donor of allowing a sample taken from his or her body to be used in research is to contribute to biomedical science for the benefit of all. Seen from this perspective, human biological material cannot be considered a marketable commodity, a point of view that is enshrined in the Civil Code under the principle of respecting the human body. However, technological advances such as high-throughput sequencing and the rise of health databases are challenging this outlook. The collection and use of biological samples and associated data have intensified in recent years due to the development of what is known as personalized medicine (Rial-Sebbag 2014). Samples taken for personalized medicine are used to produce fundamental knowledge (such as identifying biomarkers) that will enable the patient to benefit from the best treatment in the long term. Hence, this new approach to medical care requires taking large numbers of samples, which, in turn, has led to the development of standardized methods for collecting samples, while increasing costs for establishments hosting BRCs. At the same time, policies supporting data sharing (Rial-Sebbag 2016) and sample exchanges, and the implementation of plans for developing genomic medicine,[15] have helped develop a "market" capable of bringing together supply and demand. Consequently, people now commonly use the language of tariffed exchange, such as "sample price" (Clément et al. 2014), "balancing BRCs' budgets so they can ensure their operations are sustainable" (particularly through selling biological resources, while maintaining that the price is to cover the BRC's operating costs and not a way of attributing a value to samples, which is forbidden by the ban on trade in human biological material), and "contracts" (especially Material Transfer and Data Transfer Agreements), etc. To this must be added the growing demands of researchers and their institutions for recognition for their work on biological samples (via intellectual property tools and patents) and the development of new biotechnological products by pharmaceutical laboratories. This convergence of interests has generated a new form of bio-health economy (OECD 2009), combining financial support for research, the structuring of the relationships between the different operators, their interactions, and possible ways of rewarding them.

The advent of very high capacity, interoperable health databases has also helped crystallize the "market". Thanks to the "big data" movement (Rial-Sebbag 2017), it is now possible to compile data from numerous sources to increase the statistical power of analyses, for example, when attempting to identify new biomarkers or when developing new prevention strategies by better identifying groups of people who are at risk. European legislation provides a framework for operating these

[15]In France a *Plan France genomic medicine 2025* has been adopted in 2016, http://solidarites-sante.gouv.fr/systeme-de-sante-et-medico-social/recherche-et-innovation/france-genomique.

databases[16] and for handling the personal data[17] they process. This framework aims to protect the rights of both database designers and the people from whom data are obtained. Although the aim of this framework is to find a balance between the interests of database developers and the rights of individuals in the internal market, these instruments are very general and must be applied in the field of health and research (Chassang 2017). In contrast, in 2016 the World Medical Association adopted a "Declaration on Ethical Considerations for Health Databases and Biobanks"[18] (known as the Taipei Declaration), which focuses on health data used in research. Its preamble provides the following definition of health databases and biobanks: "A Health Database is a system for collecting, organizing and storing health information. A biobank is a collection of biological material and associated data. Biological material refers to a sample obtained from an individual human being, living or deceased, which can provide biological information, including genetic information, about that individual. Health Databases and Biobanks are both collections on individuals and population, and both give rise to the similar concerns about dignity, autonomy, privacy, confidentiality and discrimination." The Taipei Declaration's objective is to reconcile the interests of the actors involved and to protect donors, who are potentially the system's least powerful stakeholders. In fact, this declaration follows on from the Helsinki Declaration, which was designed to protect research subjects and therefore protect, indirectly, any resulting data However, the Taipei Declaration does not overlook the issue of a market developing, as Article 18 states: "Special considerations should be given to the possible exploitation of intellectual property. Protections for ownership of materials, rights and privileges must be considered and contractually defined before collecting and sharing the material. Intellectual property issues should be addressed in a policy, which covers the rights of all stakeholders and communicated in a transparent manner." Although these instruments were drawn up with respect to the collection of medical research data within the health system, they should also be applicable in competitive "market" situations such as those created by direct-to-consumer genetic tests (Anastasova and Rial-Sebbag 2012). The genetic testing market, in which private companies offer clients direct genetic testing via the Internet, involves storing genetic data on the Internet and is subject more to data protection and consumer protection law than it is to legislation governing medical research (CNIL 2017). Hence, this market is developing under a variety of legal frameworks, which makes donors' rights with respect to their samples and data less clear. To

[16]Directive 96/9/EC of The European Parliament and of The Council of 11 March 1996 on the legal protection of databases.

[17]Regulation (EU) 2016/679 of the European Parliament and of the Council of 27 April 2016 on the protection of natural persons with regard to the processing of personal data and on the free movement of such data, and repealing Directive 95/46/EC (General Data Protection Regulation).

[18]World Medicla Association, Declaration of Taipei / Research on Health Databases, Big Data and Biobanks, https://www.wma.net/policies-post/wma-declaration-of-taipei-on-ethical-considerations-regarding-health-databases-and-biobanks/.

address this issue, it is necessary to examine the effectiveness of laws governing the non-commercialization of the human body and property rights.

9.2.2 The Effectiveness of Laws Governing the Non-Commercialization of the Human Body and the Lack of Property Rights on Biological Samples and Data

Lawyers have long studied the notions of the ownership and non-commercialization of the human body, with even greater attention being paid to these issues since it became possible to separate parts of the body, preserve them, and use them for purposes other than grafts. These developments have raised the possibility of the human body being commoditized (Galloux 1989). The 1994 Bioethics Act and subsequent revisions (in 2004 and 2011) added specific protection of the human body, based on the principles of solidarity and integrity, to the Civil and Public Health Codes. The aim of this legislation was to rule out any possibility of human body parts being commoditized. Thus, paragraph 3 of Article 16.1 of the Civil Code states: "the human body, its parts, and its products cannot be subject to pecuniary rights". Moreover, the Civil Code clarifies the scope of this prohibition by declaring null and void any agreement conferring a financial value on the human body[19] and by forbidding the payment of any remuneration to an individual in exchange for samples from that person's body.[20] However, although these principles are considered a matter of public order, in practice they are of limited scope. First, they apply only to the person from whom samples are taken and the person who harvests these samples; there is nothing in these laws that clearly covers post-collection exchanges, except Article 1162 of the Civil Code (replaced Article 1128 on tradable goods), according to which "The contract cannot contravene public order either by its stipulations or by its purpose, whether or not this purpose is known to all the parties". Thus, any agreement, including cession, which would be subject to a tariff paid to the BRC by an applicant, may be in the form of a financial transaction, although the monetary value of the transaction is not an evaluation of the sample but of the service rendered (storage, extraction, and preparation). In this case, a donor cannot receive payment for his or her contribution because he or she does not own the sample (Le Corre et al. 2016) (or his or her body). In addition, any intellectual property rights that might be claimed on the sample can only be claimed for the researcher (inventor) or the researcher's institution. Finally, if a marketing authorization is granted to a medicinal product, the fruits of any sales may be collected only by the manufacturer and any other actors with whom collaboration contracts have been concluded.

[19]Article 16-5 Civil Code.

[20]Article 16-6 Civil Code.

In all these transactions, which are now lawful, the parts of the human body are considered totally detached from the body and are sufficiently transformed (with respect to their physiological state at the time of collection) to be the object of a financial transaction. However, the donor does not have the legal right to be involved in the commercial valuation mechanisms for samples and no compensation can be granted. Similar considerations apply to data associated with samples; because the source person does not own the data, he or she cannot claim any pecuniary rights. In both cases, donor information and consent mechanisms mean that the only rights granted are those concerning the use of samples and data.[21]

Although French law has put individuals' rights at the heart of legislation governing the use of biological samples and associated data, these rights are effective only with respect to the initial harvesting of samples and they tend to become increasingly tenuous with successive uses of the sample. Thus, health law has focused on protecting individuals' rights within the medical relationship, but it has shown its limits when it comes to recognizing these rights in recovery operations. Hence, given the predicted growth in the samples and data market, it is becoming urgent to recognize the contribution made by donors. The criteria used to associate donors with the contributions they make to science should be considered in a more global way, both when samples are collected and when they are assigned a value. In this respect, governance based on transparency is in the interests not only of the donors but of all the actors.

References

Anastasova, V., & Rial-Sebbag, E. (2012). Les tests génétiques en accès libre: quelle protection pour le consommateur européen ? *Revue de Droit Sanitaire et Social, 5*, 817–827.

Bellivier, F., & Noiville, C. (2009). *Les biobanques, Que sais-je*. Paris: PUF.

Chassang, G. (2017). The impact of the EU general data protection regulation on scientific research. *Ecancermedicalscience, 11*, 709. free access at https://www.ncbi.nlm.nih.gov/pmc/articles/PMC5243137/.

Clément, B., Yuille, M., Zaltoukal, K., Wichmann, H. E., Anton, G., Parodi, B., et al. (2014). Public biobanks: Calculation and recovery of costs; EU-US Expert Group on cost recovery in biobanks. *Science Translational Medicine, 5–6*, 261fs45.

CNIL (National Agency for Data Protection). (2017). Les données génétiques (pp. 177 et s). Paris: La documentation Française.

Galloux, J. C. (1989). Réflexions sur la catégorie des choses hors du commerce: l'exemple des éléments et des produits du corps humain en droit français. *Les Cahiers de droit, 30*(4), 1011.

INCA (National Cancer Research Institute). (2011). *Conservation and use of tumor samples in oncology*. http://www.e-cancer.fr/Expertises-et-publications/Catalogue-despublications/Conservation-et-utilisation-des-echantillons-tumoraux-en-cancerologie

Le Corre, F., Chassang, G., & Rial-Sebbag, E. (2016). Valorisation des éléments du corps humain: biobanques, proriété et valorisation. *Revue générale de droit médical, 61*, 141–157.

[21] Additional rights are granted to individuals: access, rectification, to be forgotten, portability.

Organisation for Economic Co-operation and Development (OECD). (2001). *Biological Resource Centres: underpinning the future of life sciences and biotechnology*. Paris: OECD.
Organisation for Economic Co-operation and Development (OECD). (2009). *The Bioeconomy to 2030 Designing a Policy Agenda*. Paris: OECD.
Rial-Sebbag, E. (2006). Conditions d'utilisation des échantillons biologiques d'origine humaine pour des recherches scientifiques, règles de fonctionnement des tumorothèques. *Médecine/sciences, 22*(n° spécial), 8–14.
Rial-Sebbag, E. (2009). *Implications juridiques des nouvelles formes de gouvernance en biotechnologie: l'exemple des biobanques utilisées en recherche*. Thèse de Doctorat, Université de Toulouse, Toulouse.
Rial-Sebbag, E. (2014). Personalised medicine, private medicine? Legal and public health issues. *Médecines/Sciences, 2*, 36–40.
Rial-Sebbag, E. (2016). Données de santé dans un monde hyper connecté: enjeux éthiques, forum NTIC et les maladies rares, 26 septembre 2016, Paris, compte rendu publié dans les cahiers d'Orphanet, http://www.orpha.net/orphacom/cahiers/docs/FR/XVIIforum.pdf
Rial-Sebbag, E. (2017). La gouvernance des Big Data utilisées en santé: un enjeu national et international. *Journal international de bioéthique et d'éthique des sciences, 28*(3), 41.

Chapter 10
The Development of Tumor Collections: The Interest of the European Union

Nathalie De Grove-Valdeyron

Human biological resources[1] are an essential resource for medical care as well as for new therapies. Indeed, recent research in biomedical sciences is articulated around studies of characteristics, differences and genetic peculiarities, for which these resources are essential to create a personalized medicine. The latter has been able to develop using biomarkers, i.e. molecules or overexpressed or abnormally absent proteins in some types of tumors. Research of the patient who has these specific biological markers can be predictive of a positive response to targeted therapy or the evolution of a tumor and/or its severity. This discovery has helped to develop medical programs adjusted to the peculiarities of patients and, based on identified risk factors, to adapt prevention, to diagnose at an early stage, and to propose an appropriate treatment. Beyond the therapeutic aspect, biological samples are also a vital resource to the programs of research on the biology of cancers and the detection of genetic variants. Samples are collected using a process and within very strict legal[2] and bioethical (Bioy 2016, p. 75) requirements, particularly in France. These samples are grouped into collections[3] that may have different natures: some are made for medical care (we shall not explore that aspect), others for research purposes. Conversely, a collection consists of a series of samples, that are associated

[1]What we mean by biological resources is the association of biological samples of human origin (tissue, microorganisms, cells, organs, fluids...) and related data of different types (clinical, biological, or other).

[2]See directive 2004/23/EC of 31 March 2004 on setting quality and safety standards for donating, obtaining, controlling, transforming, preserving, storing and distributing human tissue cells, JO No L 102 of 7 April 2004.

[3]Les collections de ressource biologique humaine, Repères, collection, February 2005, INSERM and Art. L.1243-3 in Code de la Santé Publique (CSP below).

N. De Grove-Valdeyron (✉)
University Toulouse 1 Capitole, Research Institute in European, International and Comparative Law (IRDEIC), Toulouse, France
e-mail: nathalie.de-grove-valdeyron@ut-capitole.fr

X. Bioy (ed.), *Public Regulation of Tumor Banks*,
https://doi.org/10.1007/978-3-319-90563-1_10

with data, re-characterized as biological resources, and representative of a homogeneous group of people or patients defined by clinical and biological criteria (Rial Sebbag 2008, pp. 233–275). These collections are managed by Centers of biological resources (CRB) now called "biobanks". These biobanks ensure conservation, and the aspect that is particularly interesting for us in this study is the availability of essential resources for the needs of the scientific community,[4] allowing their development. These biobanks are set up around a pathology, cancer. Samples obtained during sampling (often in an invasive manner), as part of the medical care provided to individuals with cancer, suspected of suffering from cancer, or susceptible to develop this disease, are then cryopreserved, following a set protocol.

The samples themselves must be of quality; they must come from a large number of people and cohorts of individuals that are clinically well defined, thus resulting in biomarkers of excellence. Banks of cells and tissues and their collections only have value through the use and exchange of biological samples and associated data and new high-speed molecular analytical techniques require a large number of samples to be able to draw meaningful conclusions.

To address these challenges in France, the Biobanks national infrastructure (Rial Sebbag 2014) which will be funded until 2020, is distributed throughout the French territory (82 centers to date). It includes biobanks but also tumor banks and microbiological resource centers (mBRCs) while involving numerous research teams (700) in projects using collections of biological samples. The infrastructure aims to increase biomedical research capacities by facilitating access to biological resources and data necessary to scientific projects of excellence.

However, biomedical research in the twenty-first century and the challenges it entails could not remain simply national and it became necessary to pool resources and expertise at the level of the European Union (Joly 2017) to facilitate the flow of information between researchers and institutions through the development of databases, incorporating a large amount of data, big data (Castets-Renard 2017).

This contribution is intended to examine the way in which the policy of the European Union promotes the development of research infrastructures devoted to biobanks, which includes tumor banks (Sect. 10.1), and to assess if the current legal framework in which they operate at European level is adapted to allow the development of tumor collections (Sect. 10.2).

[4]According to Rial-Sebbag , "biobanks are structured spaces that are designed to store, organize, use, and provide biological samples and their associated data in order to carry out scientific research (or medical care)" in *Implications juridiques des nouvelles formes de gouvernance en bioetchnologie, l'exemple des biobanques utilisées en recherche,* PhD 2009, p. 10. See also the definition in Art. 1 of the implementing Commission decision of November 22, 2013 2013-701-EU JOUE L 320 of November 30, 2013.

10.1 The Union's Policy: A Policy That Encourages Research

To appreciate the scope of action of the European Union, it is first essential to investigate its practical competence in the field of research (Sect. 10.1.1). Then, we shall consider the means at the Union's disposal to contribute towards increasing the efficiency and excellence of biomedical research (Sect. 10.1.2).

10.1.1 The Competence of the Union in the Field of Research: A Specific Shared Competence

The Lisbon Treaty has disconnected the Research, Technological Development (RTD) from the industry's short-term interests by focusing research into the European project of free movement of people and ideas (Dubouis and Blumann 2015, 2016) Article 179 §1 of the Treaty on the Functioning of the European Union (hereinafter TFEU) emphasizes their fundamental objectives: creating a European research area in which researchers, scientific knowledge and technology circulate freely; encouraging the development of its competitiveness and promoting the research activities considered necessary in respect of other sections of the Treaty. This last point may cause perplexity, since, as part of public health, research activities may also be carried out and therefore create a competition on a legal basis. The competence of the Union in the field of research is a shared competence (Art. 4§3 TFEU) but it also features specificities that are not found in other competences of this type. Indeed, the Treaty states that the Union can carry out its decisions and programs, but the exercise of this competence (Blanquet 2012) cannot prevent the States from exercising their own competence. Both competences can be exercised in parallel (Dubouis and Blumann 2016), which does not correspond to the usual principle of pre-emption which is dismissed in this case. This shared competence is unique since it dismisses the mechanics of this sharing (Blanquet 2012).

In addition, the Treaty also allows the possibility of using the open method of coordination (OMC), which is rather used as complementary and supporting competences.[5] Under Article 179§2 TFEU, "the European Union encourages the research activities of the companies but also that of universities and research centers". It can give support through direct action based on its joint research center, but it intervenes mainly in an indirect way by providing funding through a multi-year set program[6] and by advocating a broad cooperation with all the economic and scientific players.

[5]As in the field of public health, subject to the areas where the Union has shared competence in terms of common security issues [(Art. 168 §4 (a), (b) (c)].

[6]EU Regulation 1291/2013 December 11, 2013 (*JO* L 347, 20 December 2013) and decision 2013/743/CE of 3rd Dec. 2013 (specific implementation program).

The current fundraising program of the Union in the field of research is called the 'horizon 2020' program. It implements Articles 182–188 TFEU by setting the scientific and technological objectives that revolve around three priorities: scientific excellence, industrial primacy and societal challenges. Research in the biomedical field falls within the areas of eligibility of the program which provides project leaders with significant funding.[7]

Article 188 TFEU also offers opportunities that have been used to develop a European research infrastructure. According to this article, "the Union may create common undertakings or any other structure necessary for the execution of research programs". These structures are created under the ordinary legislative procedure. That is how ERIC was created (a European research infrastructure consortium) and then a few years later, a variation of it in biobanks and Biomolecular resources, *"ERIC-BBMRI.*[8]*"*

10.1.2 European Infrastructures for Research Excellence: From ERIC *to* BBMRI-ERIC

ERIC was created by Regulation (EC) n ° 723 of 25 June 2009 under the former Art. 172 and s of the TCE. It seemed necessary to "encourage the development of new structures by creating an appropriate legal framework intended to facilitate their creation and their operation on the Community scale" (recital n ° 3). This need has been formulated both by the political players (Member States and European institutions) and by the players of European research itself: companies, research centers, universities and, in particular, ESFRI (European strategy forum for research infrastructures). This new legal framework, which complements the existing legal frameworks (at a national, European or international level) provides the procedures and conditions for the establishment and exploitation of European 'Community'-wide research infrastructure necessary for the effective performance of RTD programs.

It is worth stressing that this structure brings together Member States, but not biobanks, and allows the accession of States associated with the RTD program, the accession of third States, or even of specialized intergovernmental organizations. Unlike joint ventures, the Consortium is, however, designed as a legal entity the European Union is not necessarily a member of. Moreover, the activity of an ERIC *should be non-economic*, under the terms of the regulation, "to encourage innovation and the transfer of knowledge and technology, an ERIC should be allowed to undertake *limited economic activities*, provided that they are closely linked to its main mission". These research infrastructures should also maintain the scientific excellence of European research and European research and industry competitiveness worldwide. ERIC has the broadest legal personality and capacity from the day

[7] For example, that is how colon cancer research has been funded, as part of this program.

[8] In practice, the term "BBMRI ERIC" is preferred.

the decision of creation comes into force, ERIC is financially responsible but the plan leaves considerable leeway for its members (at least three Member States) to establish the liability regime, or even change the statutes. Finally, to complete a quick statement of the main features of ERIC, it is controlled by the Commission in the form of an annual report given to the latter.

"BBMRI-ERIC" is a version of ERIC; it is a pan-European research infrastructure specifically dedicated to biobanking and biomolecular resources. This consortium, created by a decision of the Commission on November 22, 2013,[9] particularly aims at "facilitating access to resources and facilities" and "promoting quality medical and biomolecular research" (Art. 3). It should be noted that the BBMRI is guided by values (Art. 4),[10] and, more particularly, ethics, which despite various approaches in Europe, coexist in the observance of fundamental rights (Byk 2001, p. 425), and apply to all members of the consortium.

Beyond ethical issues that would themselves require a specific analysis (Burgun et al. 2016), we look specifically at the question of access to samples and data in a European or even international context, in data transfer beyond the territory of the Member States, to assess whether, and to what extent, the law of the Union can foster the development of tumor collections.

10.2 The Union's Regulation with Regards to Access to Biological Samples and Data for Research Purposes

We have stressed the particularity of biomedical research in relation to samples that are associated with health data. The latter are subject to a specific protection both in terms of the law of the Union, through the Regulation (EU) 2016/679 of April 27, 2016,[11] and at a national level, through the law of modernization of our health care system from January 26, 2016, which created the national data system,[12] which considers the recent technological developments and ensures adequate protection. We shall only consider below the rules that may apply to the ERIC BBMRI, revealing the limits of the Union's regulation (Sect. 10.2.1) before considering a few solutions that could potentially contribute to a better European governance in an effort to fit the requirements of biomedical research, to share data in a European and

[9]Implementing decision of the Commission on November 22, 2013 2013-701-EU relating to the constitution of research infrastructure dedicated to Biobanking and Biomolecular resources consortium for a European research infrastructure, OJ L 320 63 of 30 November 2013.

[10]Pan-European reach, scientific excellence, transparency, openness, adjustment capacity, ethical awareness, legality and respect for human values.

[11]Regulation (EU) 2016/679 of the European Parliament and the Council of April 27, 2016 on the protection of natural persons with respect to the processing of personal data and their free movement OJ No L119 of May 4, 2016. Applicable from May 25, 2018.

[12]See Art. L-1461-1 L 1461-7 of CSP as modified by health Act of 2016.

international context, and engage a virtuous circle for the benefit of research and ultimately patients (Sect. 10.2.2).

10.2.1 The Inherent Limitations in the European Research Infrastructure

It is necessary to point out that BBMRI ERIC, as a research infrastructure, is subject to 'ERIC' regulations which outline the legal framework and the hierarchy of legal norms. Thus, according to Article 15, the Union's regulation and particularly ERIC regulations as a secondary set of regulations, applies to the creation and the functioning of ERIC. In the hierarchy of norms, the law of the State where the head office is based comes next for questions which are not or are only partially governed by the Regulation (in this case, the Austrian law), and finally the statutes and their implementing rules. This article is stated by recital 21 which indicates that when an ERIC has an activity in another State—which is the case of the BBMRI due to its distributed structure—"other State law should apply with respect to the defined specific points in the statutes of the ERIC". This point suggests that several national laws, other than the law of the State of the head office, can apply. As seen, the legal context is broken out and supported by a spine based on the regulation of the European Union.

Among the missions of the BBMRI, Article 3 states, as already noted, that it must "facilitate access to resources and facilities, and promote quality medical and biomolecular research" [. . .] but also "contribute to increase the use and the dissemination of knowledge as well as the development of the results of the research-based biobanks across Europe". Each Member State is also required to give access to "partner biobanks, biological and bio-molecular resources and/or data of the ERIC, in respect of a *common set of standards and conditions such as defined in the charter of partnership* of the BBMRI ERIC and approved by the assembly of members" (Article 4, pt. 3, a). Finally, according to Article 18 entitled 'access', "the BBMRI puts at the disposal of researchers and research institutes, samples and data stored in databases affiliated with partners of the BBMRI ERIC, *according to the procedures and criteria approved by the assembly of members*". The regulation specifies that "none of the provisions of the statutes can be interpreted as restricting the right of owners of biobanks and biomolecular resources affiliated to the BBMRI ERIC to grant access to samples and data, whatever they may be", which is extremely significant.

These provisions illustrate the fact that the Union's regulation, through the creation of infrastructures in the biomedical and biomolecular field—the mission and role of which the Union decides on—certainly allows a pooling of resources and expertise, encourages the development of bases aggregating volume of data (big data), but it cannot harmonize (Azoulai 2015), under Art. 114 TFEU, in this area.

This lack of competence is justified both in terms of health law (Article 168 TFEU) and in terms of RTD (Art. 179 and s.).

However, the EU law on the protection of individuals with respect to the treatment of personal data aims to apply to research activity conducted by the BBMRI, no. 2016/679 April 27, 2016 EU regulation providing for a special status for health data, sensitive data. Recital 35 mentions expressly, in respect of health data, "information obtained during the test or examination of a part of the body or a bodily substance, including genetic data and biological samples" and recital 159 expressly refers to the application of Article 179 TFEU in data processing. Without further development on the regulation and the problems that arise in the context of large scale health data exchange,[13] we shall just point out that nothing opposes the transfer of health data to a Member State of the European Union as long as all Member States respect the rules and principles established by the regulation. However, in the case of data transfer outside the Union, the third State shall ensure a sufficient level of protection of the privacy and fundamental rights of the person with respect to how this data is or may be processed.

French law stipulates that sufficient protection[14] is assessed "based on the provisions in force in that State, on the security measures that are in place in that state, on the characteristics of process such as its purposes and duration, as well as its nature, origin and destination of processed data[15]". Also, in general, "treatment for purposes of research, study or assessment [. . .] are allowed according to the procedure laid down in chapter IX of law No. 78-17 of 6 January 1978".

The application of the law of the Union first, in addition to national laws and their resulting constraints—in areas not governed by the Union's regulation or for which the latter gives flexibility to states (which is the case with Regulation (EU) 2016/679)—and finally the application of the law of the head office state (Austria) leads to a "legal patchwork" whose lack of readability can be an impediment to the development of exchanges which are, however, encouraged by the Union's regulation.

10.2.2 Better European Governance as an Asset for the Development of Collections

Due to a lack of European legal framework, each national BIOBANK,[16] a BBMRI ERIC structure node, remains subject to its own law with regard to activities carried

[13]On the application of this regulation, see the communication of Carlson Fox (C), aforementioned.

[14]It is up to the European Commission and national authorities for the protection of data to decide which States meet the adequate level of protection.

[15]Art. L. No. 78-17 of 6 January 1978, Art LIL. 68, al 1 see also Art. 69, al 1 (legal exceptions) and al. 2 (contractual exemptions: "binding corporate rules (BCR). ") According to the "Privacy shield" the United States must provide "essentially equivalent" protection to that provided by European law concerning data collected within the European Union and processed on their territory.

[16]See the report on Biobanks for Europe. To avoid confusion we shall use capital letters to name the national BioBank infrastructure.

out on its territory. To facilitate a better knowledge of national law and the legal and ethical requirements applicable in France, the BIOBANK has a "common ethical service and regulation of BioBanks" (SCR), which is developing practical tools and information destined for the operators of biobanks. The website hSERN (Human Sample Exchange Regulation Navigator) or the publication of a handbook on the legal and ethical framework which is intended to apply to the activities of biobanks in health research in France (Chassang et al. 2015), is clearly the area of these support missions. This approach at a national level has also developed at a European infrastructure level, through the creation of the common "ELSI": the Ethical, Legal and Social Issues Service. To compensate for the inability to use a binding framework through harmonization, in the legal sense of the term, "soft law" is being used[17] in the field of biomedical research.

Soft law is thus used, within the European infrastructure BBMRI, to aim for "standardization" and harmonization of work practices. The regulations specifically refer to that by mentioning the development of a *partnership charter*, but it can also come in other forms such as codes of conduct—i.e. *Guidelines*. The European consortium's members have committed to defining a common policy for Biobanks, be they public or private. The recommendations and practices developed within ERIC are meant to be used as regulatory acts. For this purpose, working sessions and round tables are held regularly within the BBMRI under ELSI to find answers to sensitive questions such as those relating to consent of data use for a plurality of scientific purposes which were not foreseeable at the time of the initial consent of the individual but which have emerged from the cognitive dynamics of the research process, or the issue of the necessary recognition of researchers and clinicians at the origin of bio resources, without which a reluctance to sharing may prevail. The work plan of BBMRI for 2017[18] includes the adoption of a code of conduct (which would also be understandable for citizens) to clarify the rules on data sharing when data protection regulations will come into force. The rules of the regulations could be interpreted differently depending on each state.[19]

Doesn't the European Union have more means at its disposal to improve European governance of biobanks in general—including tumor banks—and to

[17]Here, it is worth mentioning Eurobiobank, a network in the field of rare diseases which takes part in BBMRI-ERIC and applies standard operating procedures "(SOPs) for all exchanges and samples use." This network has studied, with the assistance of independent experts, ethical issues in biobanking activities and has developed several documents available on the following website http://www.eurobiobank.org/en/documents/mta.htm. For BBMRI in relation to transfer agreement, see http://www.bbmri-eric.eu/wp-content/uploads/2016/07/LPC_MDTA_final-with-CoBRA_12-May-2016.pdf.

[18]Available online http://www.bbmri-eric.eu/wp-content/uploads/2016/07/BBMRI-ERIC_Work_Programme_2017_online.pdf.

[19]Some terms such as "pseudonymisation" or anonymisation do not have the same meaning in the UK or in Germany. These differences in comprehension may be responsible for reservations or doubts when scientists and research groups wish to share data; See article by Litton JE, available by following the link on http:// www.nature.com/news/we-must-urgently-clarify-data-sharing (January 24, 2017).

encourage the development of tumor collections? Two solutions may be considered: a recourse to the "OMC", and the development of easier access to finance.

In terms of RTD, and according to Art. 181TFEU, the Commission may "take any useful initiative to promote coordination [research between the Union and the States] including organizing the exchange of best practices". This provision, if we stick to a strict interpretation of the text, is about the coordination of research, and it is unclear how and in what capacity the Commission could be involved in the actual management of BBMRI, as it essentially intervenes as part of the activity control of the European scientific infrastructure from an operational and financial perspective through the activity report that must be submitted to it.

However, the reference to "Exchange of good practices" among the examples stated in paragraph 2 of Article 181 TFEU seems to give room for maneuver to the Commission, given the power of initiative which is attributed to it "in close partnership with the States". The report entitled "Biobanks in Europe", prepared for the Commission in 2010, was already about the need to adopt a 'harmonization'[20] in the field of samples and data storage practices and to clarify the legal context and international requirements for data sharing between different States.[21]

If the power of initiative of the Commission does not seem to be able to be achieved through a legislative proposal in this area, there is nothing to prevent the EU Executive from using the 'open method of coordination' (De la Rosa 2015) and to promote the exchange of good practices regarding the activity of European research biobanks which have a cross-border activity, or more specifically of tumor banks, and to be responsible for guidance or recommendations. But these non-binding acts are only intended to apply to Member States. The preferred and appropriate option consists in allowing scientists, researchers and lawyers, within the European infrastructure, to define a working method which clarifies the States' constraints and releases them from common requirements regarding the sharing of samples and data between several Member States and with third States, in a way that is compatible with the Union's regulation, the national laws, and in respect for human rights.

With regard to financing, the Horizon 2020 program currently seems to be the main funding tool of the European infrastructure, alongside the contributions of members. As an example, one can mention the funding secured for "ADOPT BBMRI-ERIC implementation and operation of the gateway for health into BBMI-ERIC"[22] (created in 2015 for 36 months) whose objective is to provide support to the implementation of the BBMRI and its services (ESLI is responsible for legal and ethical issues).

[20]Final Report of the joint PHOEBE—P3G—BBMRI Conference "Harmonising Biobank Research: Maximising Value – Maximising Use". 2009: Brussels cited in Biobanks in Europe—Prospects for harmonization and Networking.

[21]Biobanks report, aforementioned p. 144.

[22]See work plan 2017 aforementioned for other funding obtained under the horizon 2020 program.

Obtaining a specific Union funding for European research infrastructures in the field of biomedical research would have, in the past, required a change or, at the very least, a clarification of the ERIC regulation so that the Union could make financial contributions to the structure. If such a contribution seems theoretically possible under "other income" (Art. 5 dash 2 c, chapter II on financial provisions), Regulation ERIC-BBMRI, recital 6 of the regulations, however, specifies that an ERIC "*should*[23] be designed as a legal entity to which the community makes no financial contribution." Assuming this first obstacle has been overcome, the States—and particularly the States that are not included in such structures of research—should then agree, in this context, to grant funding which would materialize through a participation of the Union's budget, and there is no doubt that this may be the main pitfall.

The use of public-private partnerships is encouraged by the Commission (pursuant to Article 187 TFEU). They can be JTIs ("Joint Technology Initiative") which are meant to facilitate access to the market and to act as a lever for industrial competitiveness, especially in the advanced sectors (for example, the innovative drug initiative 2 "IMI2"). They can also be contractual public-private partnerships (CPPP): here, they also aim to finance some activities close to the market (e.g. the internet of the future, 5 G). Such partnerships could be involved, downstream, in the development of biomedical research.[24]

In conclusion, the European Union, as provided for in the Treaty, has a shared competence in the field of research, including biomedical research—that competence is both specific and shared—and it especially exercises a supportive role, which may seem paradoxical. This competence may seem small compared to "traditional" shared competence and, in practice, it means its actions are limited. Indeed, as previously shown, the Union intervenes mainly by supporting the creation of European research infrastructures to facilitate exchange and, incidentally—because it should not be the goal—the *economic* development of biological samples collections. Let us remember that these structures are composed of Member States and not of biobanks as such, which constitutes a limit in itself. The role of the Union is not similar in any way to a European regulation of biobanks and even the term 'European governance' of Biobanks, which have extremely varied statutes, can seem excessive in some ways. Using soft law tools (partnership charter, guidelines, code of good practice), a regulation compatible with the law of the Union and the constraints imposed by national laws gradually emerges, but within the framework of the European research infrastructure thus created.

Therefore, is the European Union involved enough with institutional players and researchers for the development of biomedical research in the twenty-first century? Will the funding secured in the context of Horizon 2020, or even beyond, be sufficient to meet the challenges of tomorrow's biomedical research? The

[23]The conditional should be highlighted.

[24]However, one should note a new trend which encourages an early hospital/researcher/industrialist partnership, e.g. in Toulouse, the role of the EDIT platform.

interference caused by the genomic era between basic research and applied research (for future clinical applications) is even greater today because of the very high throughput sequencing of the genome that allows prompt research results for the benefit of the patient,[25] and which makes it harder to distinguish between clinical standards of care and those relating to experimentation and research.

Might major ongoing projects such as the "Medical Genomics France 2025" plan that will lead to the creation of 12 high throughput sequencing platforms of the entire human genome, which should pave the way for therapeutic use, including immunotherapy which is more effective in cancer treatment, or the *Genomics England* platform which is a precursor in the field of sequencing, upset the configuration of biobanks? The structuring process of platforms at an internal level should be followed by a similar process at a European level and by a strengthening of networks which are encouraged by the European Union.

At a time when the European Union is going through a difficult period and is trying to rebuild itself,[26] biomedical research does not seem to meet its essential priorities at first glance, however, we have every reason to believe that the Union will support States involved in genomics research—which is closely related to genomic medicine—which could also have an industrial and even economic impact. Personalized medicine, a public health issue, resulting from translational research should eventually lead to a revolution in the field of care and prevention, to which the Union will give support, in accordance with the Treaty (Article 168 TFEU). Whatever the concrete future of Biobanks, in the next few years, the EU will have an important supporting role towards public authorities in search of a form of governance that does not hinder but promotes biomedical research instead, takes up the challenge of storage and the sustainability of data in the respect of ethical requirements, and ultimately to the benefit of health and general interest.

References

Azoulai, L. (2015). The complex weave of harmonization. In D. Chalmers & A. Arnull (Eds.), *The Oxford handbook of European law*. Oxford, England: Oxford University Press.

Bioy, X. (2016). *Biodroit. De la biopolitique au droit de la bioéthique* (p. 75). Paris: LGDJ. (coll. Systèmes).

Blanquet, M. (2012). fasc. *Jurisclasseur Europe: Compétences de l'Union – Exercice des compétences-Régulation.*

Burgun, A., Jannot, A. S., Rance, B., & Mamzer, M. F. (2016). Partage de données patients pour la recherche : aspects organisationnels et éthiques. *Ethics, Medicine and Public Health, 2*(3), 435–441.

---

[25]Concretely, the nature and order of the sequence of genes that make up each person's DNA are fully analyzed to determine the risk of developing cancer, to tailor treatments to the patient and anticipate his reactions to administered drugs.

[26]See the white paper on the future of Europe, Reflections and scenarios for the EU27 by 2025, COM (2017) 2025, 1st March 2017.

Byk, C. (2001). La bioéthique, le droit et la construction européenne. In *Mélanges en l'honneur de J Michaud-droit et bioéthique*, studies coordinated by H Gaumont Prat, Les Etudes hospitalières (p. 425).

Castets-Renard, C. (2017). Open data et open access des données de santé. In *Tumor Banks – régulation publique des centres de ressources biologiques en cancérologie* (colloque dir. X. Bioy, Toulouse).

Chassang, G., Cambon Thomsen, A., & Rial Sebbag, E. (2015). *Ethique et réglementation des biobanques de recherche* (ed.). Inserm, Biobanques.

De la Rosa, S. (2015). Ce que coordonner veut dire-Le développement de la méthode ouverte de coordination en matière de soins de santé. In E. Brosset (Ed.), *Droit européen et protection de la santé*. Bruxelles, Belgium: Bruylant.

Dubouis, L., & Blumann, C. (Eds.). (2015). *Droit européen matériel* (p. 444 and s). Paris: Montchrestien.

Dubouis, L., & Blumann, C. (2016). *Droit institutionnel de l'Union* (6th éd.). n° 590.

Joly, Y. (2017). Gouvernance des tumorothèques: tendances internationales. In *Tumor Banks – régulation publique des centres de ressources biologiques en cancérologie* (colloque dir. X. Bioy, Toulouse).

Rial Sebbag, E. (2008). Genèse d'un cadre règlementaire pour les collections d'échantillons biologiques humains utilisés en recherche-Exploration d'un modèle de gouvernance. *Revue Générale de droit médical*, (27), 233–271.

Rial Sebbag, E. (2009). *Implications juridiques des nouvelles formes de gouvernance en bioetchnologie, l'exemple des biobanques utilisées en recherche.* PhD (p. 10).

Rial Sebbag, E. (2014). Contribution des biobanques aux avancées biotechnologiques. *RLDI* supplément.

Reports

Decision of the Commission on November 22, 2013 2013-701-EU relating to the constitution of research infrastructure dedicated to Biobanking and Biomolecular resources consortium for a European research infrastructure, OJ L 320 63 of 30 November 2013.

EU Regulation 1291/2013 December 11, 2013 (*JO* L 347, 20 December 2013) and decision 2013/743/CE of 3rd Dec. 2013 (specific implementation program).

Les collections de ressource biologique humaine, Repères, collection, February 2005, INSERM.

Regulation (EU) 2016/679 of the European Parliament and the Council of April 27, 2016 on the protection of natural persons with respect to the processing of personal data and their free movement OJ No L119 of May 4, 2016. Applicable from May 25, 2018.

White paper on the future of Europe, Reflections and scenarios for the EU27 by 2025, COM (2017) 2025, 1st March 2017.

Work plan 2017 aforementioned for other funding obtained under the horizon 2020 program.

Part III
The Means for Tumor Collections Valorization

Chapter 11
Academic Valorization of Biobanks

Anne-Marie Duguet, Laurence Mabile, and Anne Cambon-Thomsen

Biologists and pathologists have a long tradition of collecting biological materials for use in research. To this end, they preserve residues of samples taken by health professionals for diagnosing their patients' pathologies. The resulting collections are an essential source of materials for a wide range of health- and care-related research.

Collections of biological samples have an intrinsic value which varies according to factors such as the number of samples stored, the nature of these samples, and the data associated with them. This raises the question of how collections can be enhanced and managed to increase their value. Collections intended for purely academic purposes are run on a nonprofit basis, whereas commercially managed collections are intended to make a profit for their owners by placing on the market samples private researchers can buy when they do not have their own source of research materials.

11.1 The Intrinsic Value of Collections

11.1.1 The Historical Value of Collections Held by Public Establishments

France's first legislation covering collections of biological samples for research was introduced in 2004. Prior to this, researchers at university hospitals, which combine

A.-M. Duguet (✉) · A. Cambon-Thomsen
UMR 1027 Inserm, University of Toulouse III Paul Sabatier, Toulouse, France
e-mail: anne.cambon-thomsen@univ-tlse3.fr

L. Mabile
BRIF (Bioresource Research Impact Factor), UMR 1027 Inserm, University of Toulouse III Paul Sabatier, Toulouse, France
e-mail: laurence.mabile@univ-tlse3.fr

X. Bioy (ed.), *Public Regulation of Tumor Banks*,
https://doi.org/10.1007/978-3-319-90563-1_11

teaching, treatment, and research functions, assembled collections as they saw fit to meet the needs of their research. Collections may store many different types of sample (Duguet et al. 2005, p. 213), including blood samples, residues of organ or tissue samples taken for therapeutic purposes, residues of biological fluids taken for analysis, additional samples collected during treatment, surgical specimens, proximal tissues, and autopsy specimens, etc. Such collections are perfectly legal in France and many other countries.

However, controversies still arise over the collection and use of bioresources for example, the 2001 Redfern Report[1] found that Liverpool's Alder Hey Children's Hospital, in the United Kingdom, had retained 3500 organs from deceased children without informing the children's parents, whose consent is required by law. During the resulting media scandal it emerged that England's hospitals and medical schools had retained more than 100,000 human organs. Such adverse publicity has discredited the storage of human organs and tissues although the public has no real understanding of how these materials are collected and used.

In fact, the legal and ethical questions surrounding the use of human bioresources have led to official inquiries in many countries. In France, a review of practices within certain public institutions, carried out by the General Inspectorate of Social Affairs (IGAS) and General Inspectorate of Education and Research, highlighted the need to safeguard collections of historic value and to improve storage methods, including prolonged freezing. The IGAS's report, published in 2002, included an inventory of existing collections.[2]

11.1.1.1 Census of Biological Collections

Many hospitals and universities maintain collections of biological samples, some of which are simple archives of human tissues, organs, fluids, etc., collected for medical diagnosis. Others, however, are used for educational purposes and may be exhibited in museums.

The collection at Paris' Salpêtrière Hospital, which focuses on the study of neurological pathologies, was begun in the days of Jean-Martin Charcot (1880), making it one of France's oldest collections of anatomical samples. Today, it includes 10,000 human brains. Saint-Vincent de Paul Hospital, also in Paris, is another pioneering institute with respect to collecting human biosamples. Its extensive collection of preserved fetuses is mostly used for research purposes, although it also includes numerous neonatal malformations that are no longer encountered (conjoined twins, sirens, anencephaly). Finally, the so-called "Musée de Port-Royal" at Port-Royal Hospital is a collection of women's pools, forceps, and waxes showing the different types of damage that can be caused to a fetus by

[1]The Royal Liverpool Children's Inquiry Report, www.go.uk.

[2]Report n°2002 009, Conservation d'éléments du corps humain en milieu hospitalier, March 2002, La Documentation Française, p 21 and subsequent.

using forceps. Scientists from the National Museum of Natural History carry out research on these specimens.

Most of the anatomical specimens in these collections are stored in a preserving fluid such as formalin. Other institutions have built up collections of historical artifacts, most notably medical instruments and wax models of anatomical parts made to demonstrate the effects of malformations or diseases to medical students. For example, Rouen University Hospital has set up a public museum presenting the history of surgery and obstetrics. Housed in part of the old Hôtel-Dieu, the museum includes several anatomical waxes and a calcified fetus extracted by Gustave Flaubert's[3] father, who was once the hospital's head physician.

The anatomy museum at Montpellier's Faculty of Medicine was created by a law of the Convention (Act of Frimaire 14, Year III) under which each of the country's three medical schools (Paris, Montpellier, and Strasbourg) was endowed with an "anatomical cabinet". In 1851, the conservatory's ever-growing collections were moved to a new, purpose-built home, a 60-m-long by 15-m-high gallery opposite the university's botanical gardens. The museum's ceiling is decorated with portraits of 28 famous doctors, chosen by the faculty's members in 1850, and supported by stucco columns that divide the gallery into four sections. Rows of display cases and stands present more than 5000 objects. The conservatory was opened to the public as a museum in 1945.

Paris's René Descartes (Paris V) University owns the Orfila-Rouvière anatomical collection. Like the collection in Montpellier, the original "anatomical cabinet" dates to the creation of the school of medicine just after the French Revolution. It became a museum in the middle of the nineteenth century and is now housed in the faculty of medicine in the Rue des Saints-Pères. In 1992, France's culture minister listed the museum's collection of 5802 objects as an historic monument (Act of December 31, 1913). Although the collections are now protected, they are not open to the public and can only be accessed by appointment.

As well as being used in anatomical research, these historical collections have been the subject of studies in a wide variety of fields, from anthropology to the history of art. However, they are not of interest only to academics, as laypeople also find them fascinating. Consequently, they can play a wider role in public education, which has been a key goal for French universities since the 1984 Savary Act.

11.1.1.2 The Legal Framework for Existing Collections

One of the remits of the IGAS inquiry was to examine how potential donors are given information about the collecting of biological samples to obtain their consent. Under French law, consent to take samples from a living subject covers just the act of sampling; it does not cover what the sample will be used for. In the case of post-

[3]French author, best known for his novel Madame Bovary.

mortem sampling,[4] consent is assumed unless an individual had explicitly stated that he or she did not want samples to be taken. Such refusals are recorded in a register.

The IGAS report advocated valorizing a coveted resource that is an essential tool in medical research, especially in the field of genetics, and creating a network of biological sample providers to improve the quality of the samples available. Fifteen biobanks were selected to form this network, which will raise their profiles within the research world. Existing collections must be inventoried and all public establishments must declare any collections they may have.

Two years after the report was published, France's parliament introduced a set of regulations for sample collections as part of a new law on bioethics.[5] Article L1243-3 of the Public Health Code gives an official definition of human biological samples and requires establishments to declare all sample collection programs, stating the reason(s) why samples are being conserved and describing the research program (s) for which they are intended. Under Article L1211-2 of the Public Health Code, unless they formally state otherwise, patients are considered to have given tacit consent for the repurposing of samples taken for treatment purposes so they can be used for research. Transfers of samples outside the establishment in which they were taken are subject to authorization by the Ministry of Higher Education and Research.

Before samples can be collected or transferred to other establishments, the institution concerned must apply to the committee responsible for the protection of individuals who take part in research projects and the Ministry of Higher Education and Research. This procedure is compulsory for both private and public collections.

11.1.2 The Scientific Value Added by Research

Data obtained from research based on a set of samples, and any clinical trials conducted in the light of this research, will, of course, add value to a biobank. Some research projects carry out new analyses on bioresources originally collected for other purposes, thereby enriching collections with new data and new samples.

Samples may be collected for use in either purely academic research or in commercial research. Academic research, which is carried out mostly by researchers working in public institutions, is not motivated by financial profit, in contrast with commercial research, whose results can be used for the commercial benefit of the sponsoring organization.

---

[4]Act 94-645.

[5]Act 2004-800 2004-08-06 JORF August 7, 2004.

11.1.2.1 Academic Research

Purely academic research is carried out to serve the common good, on a nonprofit basis, and without the notion of private property. Nevertheless, researchers have certain prerogatives with respect to the use and sharing of samples they have collected. Researchers who use biological samples collected in their department have a moral right over the results of their research; therefore, they have a certain amount of control over the way these results are published or disseminated to third parties.

Samples and related data collected by hospitals or public laboratories are available to the collecting institution's staff; however, biobanks and biological resource centers (BRC) may also supply outside research units with samples obtained by preserving any material remaining after the completion of analytical tests or publicly financed research programs. The collecting institution is responsible for the proper conservation, traceability, and custody of samples, as well as for protecting donors' rights, protecting data, respecting ethical frameworks, and ensuring consent is properly obtained. Links between computerized databases (epidemiological studies, cohorts, registers, etc.) allow biobanks to enrich their sample collections by incorporating data generated by research using its samples and subsequent clinical trials.

Cohorts and Registers[6]

Cohort studies, which involve collecting samples and data from a cohort of participants at intervals over an extended period, are carried out to assess the incidence and mortality rates of diseases. The *Framingham Heart Study*, for example, was launched in the United States in 1948 to identify risk factors for heart disease. Since then, it has monitored six cohorts of participants, containing a total of almost 15,000 people, and given rise to more than 2000 scientific papers. Studying these cohorts has revolutionized the understanding of cardiovascular disease.[7] The *European Prospective Investigation into Cancer and Nutrition* (EPIC) cohort study was set up in 1990 to investigate the relationship between diet, lifestyle, and environmental factors and the incidence of cancer. So far, blood samples from 521,000 participants recruited in 10 European countries have been taken and stored for future analysis. France's *Constances cohort*[8] has been designed to provide the epidemiologic research community with a wide range of health data obtained from a large cohort of participants, who undergo regular health examinations at health screening centers. Sampling and data collection are carried out in conjunction with

---

[6]Apport des cohortes à la connaissance de la santé ADSP n° 78 March 2012; www.hcsp.fr.

[7]Les cohortes au niveau international : histoire et perspectives Archan Singh-Manoux ADSP n° 78 March 2012 pp. 31–33.

[8]www.constances.fr.

three pharmaceutical companies and according to strict transparency and ethical guidelines.

Registers are lists of nominative data that are compiled on a continuous basis for use in public health research. France has 25 cancer registers,[9] including 13 general registers, 9 specialized organ registers, 2 childhood cancer registries, and a multi-center mesothelioma register. The Nordic countries have a long tradition of using registers in research. Biobanks are an additional type of register (Pukkala 2011) that are beginning to include genomic data (Brand and Proobst-Hensch 2007). Combining data associated with biobank samples with data from registers will greatly facilitate studies targeting specific sections of the population.

Publishing results not only benefits science, it also benefits researchers and their teams. Results from long-term research projects are integrated into databases that can be protected under intellectual property legislation.

Data sharing conventions set out the conditions under which samples can be exchanged, considering the principles of solidarity, equal access, quality, and cost control. Most of the projects in the 5th and 6th European Framework Programmes for Research and Innovation provide for the exchange of data and biological tissues between research centers in different European countries, which sometimes have different regulations or no legal framework. This resulted in an avalanche of recommendations and ethical rules for each project. Tubafrost3, a multinational project involving research teams in the Netherlands, France, Spain, Austria, Belgium, Italy, and the United Kingdom, was also faced with the problem of transferring biological samples between countries with differing legislation. The solution adopted in this case was to create the following coordinating rule: "if a tissue may legitimately be used for a certain kind of research in the country where it was taken and under whose jurisdiction the patient falls, it may also be used for such research in the country where it is sent to in the context of a scientific program even if in that other country other regulations would apply for research with residual tissue taken from patients under their jurisdiction" (van Veen et al. 2006).

11.1.2.2 Commercial Research

Commercial research programs may be conducted in either private or public institutions. A program's sponsor organizes the sample collection process, which is carried out under a protocol specifying the nature and quantity of the samples needed. The sponsor also finances the analysis of the samples and of the study's results. Consequently, the sponsor reserves all intellectual property rights (e.g., as the basis of patents), as well as the exclusive right to use or transfer the samples and data. The sponsor may or may not authorize the researchers to use the results for their own publications and authorizations to publish may be accompanied by restrictions,

[9] Source: www.inca.fr.

such as having to wait a certain amount of time before publishing results in scientific journals.

Sponsors of commercial research programs pay for and keep everything. The samples and data are considered goods that can be used to obtain a return on the initial investment (protection of data analysis software, patents for biological tests, etc.). In addition, sponsors are free to decide whether to publish the results for financial benefit. They can make any residual samples, research results, and follow-up data available to other researchers at a commercial rate.

For example, Eurobiobank is a public-private network of biobanks that collects and stores DNA, cells, and samples for researchers studying rare diseases. Samples are listed in a catalog showing which samples are available from which biobank. Any public or private body that collects or stores biological material relating to research into rare diseases can join the network, as long as it complies with the quality standards set out in the Eurobiobank charter.

11.2 Valorization of Collections

11.2.1 Academic Valorization Through Practice

In recent years, researchers using collections for multicentric projects have been reflecting on how collections can best be used. Calls for applications for European research programs stipulate that consortia conduct their research in accordance with European and international ethical guidelines. Nevertheless, the right to use data or collections remains vague. How should information flow and how should data be shared? How should access to existing collections and data sharing for research be organized? What is the scientific interest? What are the financial issues related to supply and demand?

11.2.1.1 Measure 1 of the "Cancer Plan": Strengthen Multidisciplinary Research

Collecting samples specifically for research is a very different process, in terms of both organization and operating costs, to collecting samples for diagnostic and treatment purposes. In the case of cancer, INCA's Biological Resources working group identified three types of tumor bank: "health" (regional tumor banks), "scientific proximity" (residual samples taken during patient care, the use of which requires updating consent, ensuring sample quantity, etc.), and "scientific" (samples collected by research teams with no institutional or geographical link with the institution hosting the tumor bank). The Advisory Committee on Biological Resources oversees the labeling of samples. Operating costs are calculated to consider the technical and pre-analytical stages of sample preparation, the prospective collection of clinical and biological annotations, and medical and scientific expertise.

11.2.1.2 The IBiSA Scientific Interest Group[10]

The "Biology, Health, and Agronomy Infrastructure" (IBiSA) scientific interest group is responsible for certifying France's BRCs and for promoting their work by steering bids to participate in international projects. IBiSA also supports projects to optimize the BRCs' databases and ensure they can interoperate with genome, phenotype, epidemiology, and clinical databases. In addition, it coordinates national policy for the certification and development of life sciences research centers and facilities, and helps these organizations obtain ISO 9001 quality certification.

IBiSA was created in May 2007, following the dissolution of the National Consortium for Genomics Research (CNRG), which had included the National Sequencing Center (CNS), National Genotyping Center (CNG), and National "Genopole" Network (RNG). IBiSA's members are France's main research and higher education institutions: CEA, CPU (University Conference of Presidents), the Ministry of Higher Education, Research, and Innovation (DGRI and DGES), CNRS, INCA, INRA, INRIA, and Inserm.

IBiSA's work certifying BRCs, begun in 2000 under the auspices of the Ministry of Higher Education, Research, and Innovation, and Inserm, has been a key factor in enabling France to create a national network of BRCs. Since 2008, this certification process has been carried out with respect to a national standard for the certification of biobanks. IBiSA's roots within the CNRG can also be seen in its work coordinating research centers, which began in the field of genomics, and in its application of a national charter. In 2011, a new body, called "Biobanques", was set up to coordinate France's biobank network. This organization is the French Node of the Biobanking and Biomolecular Resources Research Infrastructure—European Research Infrastructure Consortium (BBMRI-ERIC), set up in 2013.

11.2.1.3 Biobanking and Biomolecular Resources Research Infrastructure

BBMRI-ERIC was created to "establish, operate, and develop a pan-European distributed research infrastructure of biobanks and biomolecular resources to facilitate the access to resources as well as facilities and to support high quality biomolecular and medical research across Europe and beyond".[11] To this end, it operates on a nonprofit basis. Following a preparatory phase that began in 2008, BBMRI-ERIC officially came into being in December 2013. It is one of Europe's largest research infrastructures, as it comprises 19 member states (Austria, Belgium, Czech Republic, Estonia, Finland, France, Germany, Greece, Italy, Latvia, Malta, the Netherlands, Norway, Poland, Sweden, and the United Kingdom, plus Cyprus, Switzerland, and Turkey as observers) and one international organization (IARC/

---

[10] www.ibisa.net.

[11] http://www.bbmri-eric.eu/BBMRI-ERIC/about-us/.

WHO). To facilitate the use and exchange of samples and data across Europe it created an Ethical, Legal, and Societal Implications (ELSI) Common Service in 2015 and an Information Technology (IT) Common Service in 2016.

11.2.1.4 Public Population Project in Genomics and Society (P^3G)

The Public Population Project in Genomics and Society (P^3G) is "an international consortium dedicated to the development and management of a multi-disciplinary infrastructure that can compare and merge results from studies, biobanks, research databases and other similar health and social research infrastructures conducted around the world".[12] P^3G uses its resources to help the international research community carry out research aimed at producing more effective healthcare strategies.

National and international bodies are working to gain recognition for the value, quality, interoperability, and sharing of samples and data collections. However, they are faced with a lack of tools for measuring the impact of bioresources on research and for recognizing and rewarding the efforts needed to achieve these aims. These bodies have made remarkable contributions to the development of genomics policies.

11.3 Scientific Valorization Through Publications

The results of research carried out using bioresources, whether collected by the researchers themselves or obtained from BRCs, is mostly valorized through publications in scientific journals. Every journal is assigned an impact factor, based on its quality and prestige: the more prestigious the journal, the higher its impact factor. This ranking system can also be used to determine whether a research team is making a high impact.

Following a study of the scientific use of cohort data, Cambon-Thomsen et al. (2012) noted that few researchers clearly identify the sources of the samples used in their studies, thereby depriving these sources of recognition that would add to their scientific value. Although progress is being made in citing the sources of samples, there is, yet, no consensus on which section of a paper should give this information. Hence, it is sometimes found in the method and materials section, sometimes in the acknowledgments, and sometimes in the bibliography. Publishers are very vigilant in terms of consent and ethics, but much less so when it comes to citing the sources of samples. In addition, some collections are referred to by several different names, these names are sometimes misspelled, and acknowledgements for providing

[12] http://www.p3g.org/about-p3g/glance.

samples are often directed at individuals rather than at the institution these individuals represent.

To remedy this situation, the "Bioresource Research Impact Factor" (BRIF) initiative was launched in 2015 to develop a standardized system for citing bioresources and create an index for evaluating the research impact of individual bioresources. The BRIF international working group recognized the need to give each bioresource a numerical identifier, draw up a list of indicators, provide recommendations for citing references, and adapt regulatory frameworks for access and sharing. The guidelines they put forward for citing bioresources in scientific papers are known as CoBRA (Citation of Bioresources in Journal Articles).[13]

11.3.1 Academic Credit for Bioresources

As for all types of academic research, the only way to get credit for studies carried out using samples from a biobank or a bioresource is to publish the results in a scientific journal. As noted above, the fact that journals do not all enjoy equal status within the scientific community led to the creation of a ranking system for journals, based largely on the mean annual frequency with which recently published papers are cited by other papers. This impact factor provides a proxy for the importance of a journal within its field; however, in recent decades it has been misused as an indicator of the quality of both the work published and the researchers responsible for that work (Garfield 2006).

Given the importance of scientific journals, it is imperative that bioresources are cited systematically and consistently, and that their use is quantifiable. Currently, this is still not the case for biomedical papers.

The BRIF initiative has built on concepts initially published in 2003 (Cambon-Thomsen 2003) to devise a system that recognizes the contribution of bioresources to developing new scientific knowledge. To this end, they have developed a framework and a set of tools that can be used to obtain an objective assessment of the utilization of bioresources in research. The project brings together a wide range of stakeholders, including researchers, managers, funders, companies, and journal editors, to examine "i/ standardized identification schemes and reporting for better visibility and tracing of bioresources on the web; ii/ incentive policies from hosting institutions; iii/ rewarding actions" (Mabile et al. 2013).

An international working group, initially consisting of 134 members from 22 countries, although most members were from Europe (86) or North America (31), was created to develop the BRIF. This group was divided into sub-groups focusing on five specific themes: (1) BRIF and digital identifiers, (2) BRIF parameters, (3) BRIF in sharing policies, (4) BRIF and journal editors and (5) BRIF

[13]Bravo et al. (2015).

dissemination. This structure has evolved as the project has advanced and as other international initiatives addressing comparable issues have emerged (P3G[14] and GA4GH[15] in the fields of biomedicine and genomics; RDA,[16] ORCID,[17] and CoDATA,[18] etc. covering all disciplines). The CoBRA guidelines, which were the first step in ensuring the traceability of bioresources, were followed by the creation of the Open Journal of Bioresources (OJB), which publishes short, peer-reviewed papers that help researchers locate and cite bioresources.

11.3.2 Citing Bioresources: The CoBRA Guidelines

At present, and as described in Mabile et al. (2016), bioresources are either cited in a confusing, heterogeneous way or they are not cited at all. Bibliographic databases, such as PubMed, do not allow the use of a bioresource in a research paper to be retrieved systematically (Napolitani et al. 2016). A collaborative effort to address this problem by the BRIF journal editors subgroup, the European Association of Science Editors (EASE[19]), the EQUATOR[20] (Enhancing the QUAlity and Transparency Of health Research) network, and the bioresources research community resulted in the CoBRA guidelines (Bravo et al. 2015). These guidelines are designed to standardize the citation of bioresources in scientific papers, so their use can be traced via the Internet. Hence, CoBRA recommends mentioning in the Methods section each bioresource used in a study and citing it as an individual reference, using a standard format and, when possible, a unique identifier. Detailed recommendations are provided by the CoBRA checklist, which can be downloaded from the EQUATOR website,[21] along with most other internationally recognized reporting guidelines for biomedical research (Mabile et al. 2016).

Recent years have seen the emergence of several other initiatives aimed at improving the identification and traceability of different types of resources to facilitate open access and greater sharing (OpenAire,[22] DataCite,[23] CODATA,[5]

[14]http://www.p3g.org/.

[15]http://genomicsandhealth.org/.

[16]https://www.rd-alliance.org/.

[17]https://orcid.org/.

[18]http://www.codata.org/.

[19]http://www.ease.org.uk.

[20]http://www.equator-network.org/.

[21]http://www.equator-network.org/wp-content/uploads/2015/03/Cobra-check-list.pdf.

[22]https://www.openaire.eu/.

[23]https://www.datacite.org/.

Force 11,[24] ORCID,[4] etc.). For example, the Research Data Alliance (RDA) Working Group on Dynamic Data Citation[25] has published recommendations on making subsets of data citable. CoBRA is building connections with these groups to facilitate implementation of these recommendations and promote the use of suitable identifiers.

11.3.3 Publishing a Bioresource: A New Type of Journal

The Open Journal of Bioresources (OJB)[26] is one of a series of 'metajournals' published by Ubiquity Press to "enhance the reuse potential and impact of research"[27] and to improve access to research resources such as data and software. The idea behind the OJB is to provide a permanent paper record that recognizes the contribution to science of the people who create and manage bioresources. The information provided by the OJB enables users to cite bioresources they have accessed or referred to. The best way to do this is by including the bioresource in the traditional vehicle for obtaining academic credit: a paper in a peer-reviewed journal article. With this system, bioresources can be cited in the same way as any other journal, a process that is made easier by assigning a unique digital object identifier (DOI) to each article. Hence, each paper in the OJB acts as a permanent marker for a bioresource and follows the usual standards for citing research.

The peer-reviewed bioresource papers published by the OJB provide accurate, structured overviews and metadata for bioresources. Papers are written in a standard format which includes a description of the bioresource, the way(s) in which samples are preserved, the methods used to create the resource, and how it can be accessed (Mabile et al. 2016).

Decisions on whether to fund research projects are strongly linked to academic credit and recognition. Consequently, it is essential to find a way to ensure that new tools involved in the production of knowledge, such as biobanks, receive the recognition they deserve.

References

Brand, A. M., & Proobst-Hensch, N. M. (2007). Biobanking for epidemiological research and public health. *Pathology, 74*, 227–238.

Bravo, E., Calzolari, A., De Castro, P., Mabile, L., Napolitani, F., Rossi, A. M., et al. (2015). Developing a guideline to standardize the citation of bioresources in journal articles (CoBRA). *BMC Medicine, 13*(33), 1–12.

[24]https://www.force11.org/.

[25]https://www.rd-alliance.org/groups/data-citation-wg.html.

[26]http://openbioresources.metajnl.com/.

[27]https://www.ubiquitypress.com/site/about/.

Cambon-Thomsen, A. (2003). Assessing the impact of biobanks. *Nature Genetics, 34*, 25–26.
Cambon-Thomsen, A., Mabile, L., & Rial-sebbag, E. (2012). Aspects éthiques et valorisation scientifique des cohortes adsp n°78 March 2012.
Duguet, A. M., Martrille, L., & Boucly, G. (2005). Le corps morcelé et les restes humains. In *Le respect du corps humain pendant la vie et après la mort* (pp. 213–223). Les Etudes Hospitalières. www.leh.fr.
Garfield, E. (2006). The history and meaning of the journal impact factor. *JAMA, 295*(1), 90–93.
Mabile, L., Dalgleish, R., Thorisson, G. A., Deschênes, M., Hewitt, R., Carpenter, J., et al. (2013). Quantifying the use of bioresources for promoting their sharing in scientific research. *Gigascience, 2*(1), 7.
Mabile, L., De Castro, P., Bravo, E., Parodi, B., Thomsen, M., Moore, S., et al. (2016). Towards new tools for bioresource use and sharing. *Information Services and Use*. https://doi.org/10.3233/isu-160811.
Napolitani, F., Calzolari, A., Cambon-Thomsen, A., Mabile, L., Rossi, A. M., De Castro, P., et al. (2016). Biobankers: Treat the poison of invisibility with CoBRA, a systematic way of citing bioresources in journal articles. *Biopreservation and Biobanking*. (Brief report).
Pukkala, E. (2011). Biobanks and registers in epidemiologic research on cancer. *Methods in Molecular Biology, 675*, 127–164.
van Veen, E. B., Riegman, P. H., Dinjens, W. N., Lam, K. H., Oomen, M. H., Spatz, A., et al. (2006). Tubafrost3: Regulatory and ethical issues on the exchange of residual tissue for research across Europe. *European Journal of Cancer, 42*, 2914–2923.

Chapter 12
Commercialization Through the Use of Private Law Contracts

Laurent Izac

12.1 Introduction

The possibility to organize the commercialization of biobanks using private law contracts may seem doubly surprising.

Indeed, each of the two themes included in this proposal— commercialization and private law—seem to go against the first intuition of biobanks to be removed from the temptation of commodification inherent in things of economic value and that, because of their nature and of their particular vocation, they escape private law for the benefit of public law.

However, this first intuition is immediately contradicted by the finding that there is already some contractual practice providing a form of marketing of biobanks. Just consider the conventions of making available all or part of certain collections. There are sometimes real cessions for the benefit of third parties that are not necessarily public people.

Moreover, in a context of globalization of scientific research, and to organize international networking,[1] the French and European regulation frame the conditions in which biobank imports or exports can happen.[2]

The initial astonishment should therefore give way to a recognition of reality. Intuition should give way to pragmatism.

[1]E.g. BBMRI-ERIC: Biobanking and Biomolecular Resources Research Infrastructure—European Research Infrastructure Consortium.

[2]Art. R. 1235-7 to R. 1235-11 of Public Health Code.

L. Izac (✉)
University Toulouse 1 Capitole, IEJUC, Toulouse, France
e-mail: laurent.izac@univ-tlse1.fr

X. Bioy (ed.), *Public Regulation of Tumor Banks*,
https://doi.org/10.1007/978-3-319-90563-1_12

Yet, this first intuition deserves some attention. Indeed, an analysis seems to suggest that, legally,[3] the commercialization of biobanks through private law contracts is not as obvious.

The marketing aspect is the first questionable issue. According to the principle of extra-patrimony of the human body,[4] the 'marketing' of its elements seems impossible. In fact, the very term generates some embarrassment in scientific literature on the issue of the exchanges of biological samples and personal data for research purposes. So, when one refers to the fact that "*these exchanges can be granted for free or as a commodity*", they immediately specify that "*the commercial character of an exchange of biological resources should not be confused with the sale of the resource itself*" and that, therefore, "*the term "commodity" means the payment of management fees and the preparation or delivery of the resource, these operations being costly and guaranteeing the quality of the biological resource which gives it scientific value*"(Chassang et al. 2016, p. 107).

Also, as part of a more general use, one should probably use the term "development" ("valorisation" in French) of the collection instead of the word " commercialization".

But it is true that the French economic term can present a certain ambivalence because it sometimes means the process of determining the value of an object (it is close to the notion of assessment, even if, unlike the latter, it is about a process which will give a value to something that did not initially have any) and, in other cases, the process to improve the value of the object and, incidentally, the resources it may provide.

However, in a more synthetic way, the French term "valorisation" means the "*development of something to get more resources*".[5] It is exactly this process we are referring to since collections cannot remain unused and must accordingly be developed.[6] This development of the collection can certainly go through direct operation. Thus, the administrator themselves would lead research on the collection. But this development of the collection can also take the form of a provision for the benefit of other research organizations; pluralism in research often means added value. Yet, this provision, and more broadly the circulation of collections, implies some availability for the Biological Resources Centre (BRC) wishing to contract with a third party to give them access to the collection. In strict legal terms, and from the point of view of private law, the marketability of the collection must be examined.

---

[3]At least, in French law, on which we shall focus our comments.

[4]Art. 16-1 of Civil Code.

[5]According to the definition given by the Centre National de Ressources Textuelles et Lexicales.

[6]We are closer here to the objectives set by Article L.111-1 of the code of research:

> The national policy of research and technological development aims to:
>
> (1) increase knowledge;
> (2) share the scientific, technical and industrial culture;
> (3) develop the results of research to the benefit of society [...]

It is then the possibility of resorting to private law which should be discussed. Should the conclusion of a contract governing the 'marketing' of a biobank be a matter of private or public law?

One should remember that an administrative agreement may be identified in two ways.

Firstly, it is the law that has the authority to decide how to label any particular contract under the banner of public law. This is particularly the case for public-private partnership agreements regulated by the order of 17th June 2004. Under that law, such contract is automatically an administrative contract. However, regarding the development of biobanks, practice commonly resorts to this terminology when, with a concluded agreement between a biobank and a private person, it evokes a "*public-private partnership*" (Chassang et al. 2016, p. 174). If one follows the terminology used, one would then conclude to a simple eviction of private law.

However, it is not the case because this language use is misleading. Indeed, the convention covered by the ordinance of 2004 covers a scope of activity different from that which interests us. It is in its broadest sense that the expression used in practice should be understood: an agreement between a public and a private person.

This context remains conducive to public law competence.

Second, in the absence of provision in the statutes, administrative case law has reached a double standard of qualification for public contracts. The first so-called "organic criterion" requires the presence of a co-contracting public person. However, in practice, the vast majority of BRCs are managed by public persons.[7,8] The second criterion called "material" is twofold since it consists either in the execution of a public service mission or in the presence within the contract of a system beyond the scope conferred by common law. On the first aspect, conducting a common biological research in human health could be likened to the exercise of a public health mission. But it is also true that the convention may simply organize making all or part of a collection available. On the second aspect, it may happen that the agreement with the private partner contains several extremely strict clauses, particularly to control research conditions and the absolute respect of its purpose. Can it be considered as the presence of a system beyond the scope of common law because of the importance of constraints imposed on the private partner[9]? The mere fact that the question can be asked shows that the competence of private law in the marketing of a

---

[7]For a list of BRCs affiliated to the *Biobank* network (with 83 BRCs representing more than 700 research teams), see the following website: http://www.biobanques.eu/fr/annuaire-des-biobanques.

[8]It is also possible and common that a convention binds two public people, in which case, the jurisprudence of the Court of conflicts (TC, 21st March 1983, PSU) assumes that the contract falls under public law. It is however only a mere presumption that the material criterion can reverse.

[9]According to the jurisprudence of the Court of conflicts (TC, 13th October 2014, Axa company), two cumulative conditions are necessary to consider that a clause is beyond the scope of common law: it must necessarily be exercised for general interest purposes and either give "*the public person prerogatives or advantages beyond the scope of common law*" or impose "*to the other party obligations or constraints beyond the scope of common law*".

biobank is not necessarily obvious. But even when this is the case, the legal tools provided by private law should be chosen with care.

Thus, ultimately, the commercialization of a biobank using private law contracts leads to adjudication on two successive points:

First, the marketability of the object (Sect. 12.2).

Then, the efficiency of the legal instrument used (Sect. 12.3).

12.2 The Marketability of the Object

Is the marketability of the collection possible in the light of the requirements of private law? Two further reasons explain that the question deserves to have been previously decided: on the one hand, the recognition of the collection as an asset is discussed (Sect. 12.2.1), on the other hand, French public order brings a certain hindrance to its availability (Sect. 12.2.2).

12.2.1 A Complex Object, the Asset Nature of Which Is Discussed

First, it is essential to inquire as to which legal qualification should be attributed to a collection. The issue is of major interest since it is only if the collection can be qualified as an asset that its eventual marketing could be considered.

The possibility to legally mobilize something presupposes that it has previously been appropriated. In this respect, Art. 537 of the Civil Code lays down the principle of free disposition of property, i.e. appropriated things. As is well known, this same article provides the existence of a special regime regarding property owned by public entities. But it comes down again to public property, and so assets.

For private law, the recognition of the asset nature of a collection is facing two challenges. The first difficulty lies in the possibility of appropriation of the collection.

The prohibition contained in Art. 16(1) of the Civil Code provides a significant barrier. The ban to constitute a patrimony right on the human body or one of its elements seems to directly oppose all accession of the legal quality of asset for collections. However, it is essential that the law empowers science to make progress in human health. And, as shows the context in which the collections were originally formed, this objective of general interest requires a legal regime that is adapted to the needs of medical research to allow the implementation of a network and so the circulation of all or parts of collections.

However, the extra-patrimony of the body combined with the special mission of the collections may cast doubt on the relevance of real ownership or, at least, on a regular appropriation. Thus, following Art. 714 of the Civil Code,[10] some authors (Rochfeld 2014, from p. 351) have been able to propose the use of the notion of "common goods". This concept allows one to account for the rise of concerns surrounding the start of a resource being used by a community of people, for the way in which it will be preserved for the benefit of this community, or even for the resource to pass on to future generations. Legally, it provides the reservation of the collection using a mechanism with the double advantage to submit to a legal regime appropriate and common to public and private individuals.

Whatever is the consecration of this notion, as long as the law gives a collection manager the right to refuse access to third parties or not, it seems difficult to deny this collection the quality of asset. Indeed, this power to exclude is the paramount specificity of a legal reservation and, therefore, of an assumption of ownership.

The second difficulty lies in the loopholes of the statutes defining the collection as a legal object. Indeed, Art. L. 1243-3 of the Public Health Code on the conditions of collection creation, does not perfectly account for the complexity of the elements that compose it: it only envisages the physical side of it.[11] However, the composition of a collection presents a strong heterogeneity. It is true that it includes biological materials, but also medical and personal data, as well as a scientific expertise which is somehow used to annotate these materials. The intangible part of the collection does not feature in the definition of the Public Health Code. It is probably because of this failing in the legal definition of a collection for why it is often referred to as a "biobank".[12]

Whatever the terminology used, and despite the shortcomings of the statutes, the law enables one to grasp the complexity of a collection. This means referring to the legal notion of universality. Universality is the notion which enables one to legally account for bringing together—which can be more or less complex—a set of things that has a certain heterogeneity or not. The unifying element of this set is the common assignment of its elements. Thus, universality is independent from the elements it contains. Conversely, the assignment of the collection gives it its own legal existence and identity. In terms of collection, the content of this assignment is imposed by law since, according to Art. L. 1243-3 of the Public Health Code, the collection may not be created only for "*scientific purposes*."

[10]This statute aims, more specifically, at common things which it defines as the "*things that do not belong to anyone and whose usage is common to all*". The change from the category of common thing to that of common asset, therefore, requires its ownership.

[11]Art. L. 1243-3. Three of the Public Health Code al. two aims at "*gathering, for scientific purposes, biological samples taken from a group of people identified and selected based on the clinical or biological characteristics of one or more members of the group, as well as derivatives of these. Samples*" .

[12]This expression is more comprehensive, but it is not fully satisfactory because of the metonomy effect it may convey (see below).

However, although this legal assignment gives the collection its coherence and usefully puts it under an administration unit,[13] it also widely affects its marketability.

12.2.2 A Complex Object, the Marketability of Which Is Limited

The marketability of an asset defines its ability to be the object of a voluntary legal operation. It allows one to legally state whether the asset concerned is fit to be intended for the benefit of a third party. Thus, marketability is the counterpart in the order of things that capacity is in the order of people. The marketability of things and the capacity of people combine to determine whether to enter into a legal act on a thing. This legal act will only be valid on the double condition that the contractual parties have the required capacity and that the thing which is the object has the required marketability.

Although traditionally marketability and extra-marketability are opposed according to a binary distinction, marketability comprises different levels to meet the gradation of solemnity that is also part of legal acts. Just as there are acts of conservation, administration or disposition, the marketability of an asset can be more or less pronounced to make it fit or not for one of the proposed legal acts. From this point of view, the ownership of an asset offers the highest level of marketability as it makes it capable of the most solemn acts: acts of disposition, i.e. those providing the legal circulation of an asset for the benefit of a new owner. Conversely, according to a more economic approach, the ownership of an asset turns it into a commodity.

Also, it is hardly surprising that Art. 16(1) of the Civil Code establishes the principle of the extra-patrimony of the body[14] to protect it from the temptation of its eventual commodification. It should be noted, however, that under the very terms of the Civil Code it is matter of extra-patrimony and not extra-marketability. This precision accredits the idea that the body remains an asset on which the person can act legally with some flexibility: to derive revenue through its labor force[15]; or for its preservation needs.[16] This is the case especially during surgery when a tumor is extracted from the patient's body.

But, in terms of collections, this extra-patrimony of the body and its components is a restriction on their development. Indeed, the direct exploitation of the collection

[13]That exercised by the collection manager.

[14]On the issue of extra-patrimony for the patient, see S. Paricard's article.

[15]We think of an employment contract under which the person puts their physical and/or intellectual strength at the service of their employer.

[16]We think of a medical contract. Although, since the application of Article l. 1142-1-1 of the Public Health Code, the responsibility of the physician has become legal (to transcend the distinction between civil and administrative liability), and consent to care naturally remains a fundamental principle. It is the expression of will of the patient so that intervention on their body becomes possible. It would be difficult not to see a legal act concerning the patient's own body.

is already an important development, but the ability to make it available or sell all or part of it to a third party would increase its scientific—and also economic—potential. For this reason, the Public Health Code has organized a system whereby items and products of the human body become part of a patrimony for medical and scientific needs. Ownership of these 'human things' is provided by Art. L. 1241-1 of the Public Health Code while Art. L. 1243-4 of the Code regulates the build-up of collections likely to be transferred to a third party. Thus, this last article authorizes the collection to become part of a patrimony with an administrative system.[17]

However, the ownership of a collection and its process of becoming part of a patrimony are expressly conditioned by its scientific assignment. This assignment that gives the collection its coherence—making it a universality—also limits its marketability. Indeed, the existence and respect of the assignment of the collection being legislated, the BRC can freely call it into question and find itself in a situation comparable to that of a trustee. Article L. 1243-4 of the Public Health Code provides such a relative system whereby collections become part of a patrimony because a collection governed by the BRC is a trust patrimony.[18] Conversely, if the BRC may act on the collection and eventually dispose of it, it is only in strict compliance with this scientific assignment. Any legal or material act on the collection should be limited to its scientific purpose.

The result is that temporary provision of all or part of the collection, or even its eventual cession, can only be validly concluded in accordance with this purpose. Thereafter, that same purpose will also be an imperative for the third-party benefiting from the provision or cession, who will become in turn holder or owner of the trust patrimony of the collection.

And finally, a particular difficulty arises because of the material impact of the exploitation of the bodily elements being part of the collection: their gradual wear.[19] Indeed, even if today's technology allows fragmentation of the same sample to delay its exhaustion while a certain fungibility may exist between samples,[20] this is unfortunately not the case for all samples; some are rare, which limits possible operations. This inevitable phenomenon of wear[21] of samples during their scientific

[17]On the issue of when becoming part of a patrimony occurs, see M.-X. Catto's article.

[18]The common law of trust is regulated in Articles 2011 and those following of the Civil Code. Given its nature and its particular purpose, the collection called for a special regime of trust which should transcend the opposition between private and public property.

[19]Wear can be defined as the phenomenon of "*alteration, deterioration of a material thing, resulting from prolonged use or mechanical or chemical action*", CNRTL.

[20]Scientists estimate that the same cancers of the same organ allow identical studies.

[21]It seems useful to note the fact that this phenomenon does not make the samples part of the category of consumable things. Indeed, a thing can be consumed if the very principle of it being enjoyed lies in its destruction: this is the *desired* effect. So, fuel is a consumable thing because its use is very precisely to seek its destruction-combustion. For other goods, on the other hand, if an alteration/destruction is likely to occur when using them, it is a *sustained* effect, which is the case for biological samples.

exploitation better explains the importance of choosing the contracting partner who will be granted access to the collection.

In any event, to effectively organize the corresponding development, a decision must be made regarding the legal tool to be used.

12.3 The Efficiency of the Chosen Legal Tool

What conditions might underpin the effective organization of the collection marketability? The answer to this question depends, on the one hand, on the capacity of the parties (Sect. 12.3.1) and, on the other hand, on the choice of the terms of the contract (Sect. 12.3.2).

12.3.1 The Capacities of the Parties

For a contract to be valid, the parties must both be legally capable.

But this sole validity of the contract cannot be sufficient if one also wants contractual efficiency. Meaning, the contract should give allowance to organize the best means for the expected scientific outcomes. Therefore, an operational capacity must be added to the strict legal one: a technical and scientific capacity.

This dual capacity—legal and operational—must naturally be present in each of the contracting partners: the collection manager and the third party the contract will be agreed with.

With regard to the collection manager, i.e. of the BRC, their technical and scientific capacity will not pose any difficulty. Indeed, the possibility of building up a collection is subject to an administrative review procedure where that capacity is assessed. Articles L. 1243-3 and L. 1244-4 of the Public Health Code organize the terms of a special administrative police submitting the build-up of a collection to a regime of declaration or authorization.[22] That last procedure, being more binding, is directly linked to the assumptions of collection marketing. Indeed, an administrative authorization is required for building up a collection of tissues and cells from the human body "*for their transfer as part of a business activity'* or even' *for their transfer for free.*" In addition to this administrative control, the BRC will be a

[22]According to INCA reports:

Nearly 90% of BRCs have made a declaration of use of biological resources for their own research projects.

Sixty percent state an authorization number for the transfer of samples to another agency,

About 20% of BRCs have an authorization number for import and export of organs, tissues and their derivatives and cells from the human body for scientific purposes.

member of the *Biobank* network,[23] which will imply two requirements. On the one hand, the BRC will have to adhere to the code of ethics of the *Biobank* network. On the other hand, it will have to obtain a certification[24] attesting to its technical and scientific capacities.

However, the legal capacity of the collection manager can be a source of difficulties. Specifically, two conditions must be controlled: the legal personality of the biobank as a party to the convention and the power of the signatory to engage it. First, the basic status of legal personality must absolutely be met before the convention becomes valid. Yet, whilst this legal personality seems to be generally present, it is not always the case because the term 'biobank'—commonly used to describe a BRC with a research purpose—encompasses an ambivalent legal notion. Indeed, some biobanks have a legal personality, but others do not: they are the result of a simple contractual arrangement combining public and private people (Chassang et al. 2016, p. 48). Moreover, one should also note the metonymy in the use of the term "biobank". That figure of speech, which designates the container (the hosting and management structure of the collection) by naming the content (the collection itself), and vice versa, can cause significant confusion. Without a doubt, what forms the content—the collection—does exist, however it is not clear whether the entity that is supposed to host it has legal personality.[25] Therefore, to clear up ambiguities on this subject, the entity providing the collection should be checked for legal personality.

Then, the power of the signatory should be checked. Although the organization of biobanks varies from one to the other, quite often they have a scientific manager and an administrative manager. The latter should have the delegation of signature or power to validly engage the biobank. This requirement will be even more important that the convention will not only organize a simple temporary provision, but also an actual transfer of all or part of the collection. Finally, the rules of operation of the public person may demand that the contract is also signed by the financial control board.

Regarding the receiving contractual partner of the provision—usually an industrialist—legal capacity requirements are usually met.

Indeed, the industrial partner is a legal person in private law to whom the system of 'legal powers' of the manager will provide a strong legal security. If the latter is

---

[23]The *biobank* network—which has no legal personality and is hosted by a service department of INSERM—includes 83 BRCs. That represents more than 700 research teams.

[24]This certification corresponds to standard AFNOR NF S 96-900 *"Système de management d'un Centre de ressources biologiques (CRB) et qualité des ressources biologiques d'origine humaine"* published in 2008 and/or standard ISO 9001. For more information, see Mr. J.-H. Di Donato's article.

[25]We find this ambiguity in the terminology used by the European Commission when it comes to "*biobanks*": they will refer to *"collection"* or "*storage and distribution centers"*. Decision of the Commission implementing no. 2013/701UE (JOUE L 320/63) of November 22nd, 2013 on the Constitution of Research Infrastructure dedicated to biobanking and biomolecular resources consortium for a European research infrastructure (ERIC-BBMRI).

not a signatory, one needs to ensure that the mandate received by the representative of the legal person is sufficient to validly engage them. The legal personality of the industrialist will not be an issue, due to the registration requirement with the RCS demanded by French law. In the presence of a foreign company, one should remember that, in the absence of an international treaty governing the matter, the condition of reciprocity must be respected: the recognition of the legal personality as well as the extent of the legal capacity in the eyes of French law will depend on the reciprocity for a French company facing the law governing the foreign company.

As for the contractual partner's scientific and technical capacity, it deserves more attention. If, as we mentioned, joining the ethical charter of the *Biobank* network, as well as obtaining a certification, apply systematically to the BRC, it is not apparently the case for an industrial partner while the collection convention is not subject to a specific administrative control.[26] However, it is both logical and necessary that this double requirement should also apply to the contractual partner of the biobank (Chassang et al. 2016[27]). As the purpose for which the collection has been built up must be respected by the industrial partner, the same applies to the means[28] at their disposal and how they will use them to present the same level of requirement other than that expected of the BRC. The particular nature of the collection urges it because, beyond the scope of operations authorized by the scientific purpose of the collection, a qualitative requirement must also be expected from the industrial partner. This level of quality must be all the greater in that the biological resource to which they have been given access is rare and, therefore, valuable. Conversely, the exploitation of the collection should not only be allowed by the convention,[29] it should also be imposed on the industrialist, according to strict quality standards. It is only on this double condition that the convention will truly organize the development of the collection.

Finally, in the event of the industrialist breaking their commitments and, thus, engaging their liability, their financial capacity would be checked.[30] Following this line of thinking, the provision of a guarantee or the use of insurance[31]—which would

[26]Its conclusion is evoked, if at all, in the authorization process provided for by Art. L. 1244-4 of the Public Health Code. Art. R. 1243-4 6 ° of the same code, which specifies the application procedure, indicates that it will comprise the list: *'(d) if applicable, [of] manufacturers of medical devices or pharmaceutical companies that will be the beneficiaries of transferred tissues or their derivatives*'.

[27]Annexes 3 and 4 suggests that the code of ethics of the *Biobank* network should be annexed to the convention.

[28]The use of contractors by the industrialist should be subject to a special scrutiny. It would be good practice to clarify whether it is allowed or not. In the latter case, there should be an indication, on the one hand, of how it can be used and, on the other hand, of how the industrialist should supervise and control the work done by the subcontractor.

[29]In accordance with the purpose of the collection.

[30]It is possible that, for tax or financial reasons, the industrialist take part in the convention through one of their subsidiaries, the solvency of which would be disproportionate to their own.

[31]This insurance could be a damage insurance and/or a liability insurance.

be the responsibility of the industrial partner—covering the damage caused by the loss or misuse of the collection could be considered.

In any event, the conditions of provision of the collection, as well as the diligence expected from the parties or what the results are destined for, will have to be specified in the terms of the contract.

12.3.2 The Terms of the Contract

The effectiveness of the legal tool used is measured by the deployed resources and objectives in mind. In this respect, the needs of the parties can be very variable. Sometimes it will be necessary to organize an extensive scientific cooperation requiring the implementation of a sustainable governance structure. Other times, on the contrary, the convention will provide a simple transfer of a part of samples from the collection. In other cases, the convention will organize a one-off collaboration.

Therefore, everything will be a matter of choice for the parties. But, whatever their choice, it will have to be made whilst making sure to determine the most appropriate legal instrument. Three key points should be addressed: the precise definition of the object of the contract, the conditions of its execution, and what the results are destined for.

First, considering the object of the contract, it is the general structure of the proposed transaction which will serve as a guide. As we have outlined it, the agreement may vary from the one-time transfer of a fraction of samples to the organization of a sustainable partnership, but also to ad hoc collaboration. It is in the light of these three positions that the general structure of a contract may vary.[32]

In terms of the law, the main difficulty lies in the fact that these choices are not necessarily translated into agreements known to the law and benefiting, consequently, of a predefined legal regime—the so-called named contracts. To name these different types of choices, it is, therefore, not surprising that generic terms such as 'making available', 'collaboration' or 'consortium' are used in practice. These terms do not specifically refer to a convention variety already organized by law.

If we look, for example, at the first term—'making available'—it may correspond to two different mechanisms: a temporary availability—which would be close to a sort of loan or rental—or an actual transfer—a sort of sale the lucrative dimension of which remains uncertain or at least very variable. Meanwhile, the term 'consortium' refers to the constitution of a group without a legal personality; however, we cannot determine in advance if it will eventually come under a joint venture or an original convention (Chenut 2003[33]). Nevertheless, the governance of the consortium must be organized because the sustainability of the partnership implies a "management" body.

[32]Given that intermediate positions will remain possible.

[33]The consortium is considered in doctrine as being the intermediate level between indivision and society. In the absence of a legal definition, it is generally considered as a cooperation agreement,

Ultimately, in the absence of special contracts directly meeting the needs of the BRC and its contractual partner, today it is impossible to find in contract private law a legal tool that would be ready to use. Instead, the contractual tool should be forged by a smart use of law resources. But the originality of the convention thus developed will often have to borrow, at least in part, to the field of *sui generis*. However, if allowed by the conventions, the exercise requires both caution—i.e. not violating public order rules which remain applicable even in the presence of an unnamed convention—and precision—i.e. making clear and giving details as to the purpose of the convention, because law will not easily fill the gaps in the agreement. A specific standardization of the conventions could be a valuable tool which would facilitate contractual negotiation while contributing to legal certainty.

Thereafter, for the execution of the contract, the expected diligence of each party should be determined, as well as the sanctions applicable in case of breach.

With respect to the obligations of the parties, practice commonly resorts to the drafting of a Material Transfer Agreement. An MTA will state the commitments from the parties as to the provision of the collection. In addition to the constitution of a possible governing body, the financial terms, the respect of administrative formalities and the naming of the responsible partner and the purpose of authorized research, two issues seem worthy of our attention: the traceability of transferred elements and their responsibility.

The traceability of transferred elements is an important requirement to ensure their preservation and proper use. The system of fiduciary property previously referred to calls for that. Transport, storage and use of transferred elements must be provided within appropriate conditions. They should be more stringent if the material provided is rare.

The responsibility of transferred elements will proceed from a transfer of custody.[34] The contractual partner shall ensure the good preservation of the samples provided, but will also assume the risk of possible damage caused by the latter. Handling the risks inherent in the use of the collection may be stepped up by a warranty disclaimer to deprive the contractual partner of any recourse action against the BRC provider. Such a disclaimer is perfectly valid between professionals, and more so when the transferred material has an experimental nature, the quality and future behavior of which are partly unpredictable. According to the jurisprudence, this disclaimer would, however, be invalidated if the BRC was guilty of gross misconduct.[35] However, it should be noted that, during the transfer of organic material as in any economic operation, the price remains the counterparty for the

more commonly called a joint venture, concluded between two or more physical and/or moral persons committed to mandate, each with respect to themselves, the distinct outputs to a specific joint operation.

[34]According to jurisprudence, for this transfer of custody to be legally effective, the one who receives the thing should have ways to prevent the damage it is likely to cause. The technical and scientific capacity previously mentioned will also be essential here.

[35]One can think, for example, of serious failure in terms of certain technical rules.

service but also for the guarantees that come with it. Also, the contractual deprivation of any quality guarantee regarding the elements provided cannot be absolute in the case of provision for consideration. On the one hand, a number of obligations of means should weigh on the BRC to participate as well as can be in the provision of quality material.[36] On the other hand, a minimum level of quality will probably eventually emerge with the development of a collections 'market'.[37]

Concerning sanctions of any failure, the convention should *at least* provide for the consequences of failure to comply with the purpose of the collection or its faulty loss.

Failure to comply with the purpose of the collection should be expressly sanctioned by the forfeiture of the right of use granted to the contractual partner; this clause of forfeiture should be applicable even in a cession case. In effect, the system of fiduciary ownership requires respect for its object or the latter may immediately return to the sender.

The wrongful loss of a collection should also be sanctioned by the convention which could usefully organize it through a penalty clause. This loss should be understood in a very broad sense, as being outright loss by destruction, as well as a diversion in the use of biological materials made available. Indeed, in the latter case, part of the biological resources would have been used contrary to its purpose and would, therefore, have been illegally exhausted.

Finally, with regard to the fate of the outcome of the convention, we can identify two types of benefits: the scientific benefits and economic benefits.

For the former, it may be possible for the BRC to have access to a number of scientific results from the exploitation of the collection, which would be intended to be integrated into its own research program. In return, the origin of the results must then be mentioned to identify the role of the industrial partner. Depending on the nature of the convention, these results can also give rise to a joint publication involving representatives from each of the partners.

For the latter, it is essentially the issue of patents which will have to be organized by the convention. Generally, in case of mere provision, the industrial partner will keep the economic benefit of the results obtained and thus the corresponding possible patent. But a different agreement can still be reached. In case of further partnership, a process of results pooling will often be used. It may vary from a system of co-ownership of the patent to its exclusive allocation to one of the parties but with an operating license in favor of the other party.

In any event, these attribution rules will be part of the financial equation of the convention. Indeed, the criterion of return on investment will be logically determinant for the industrialist.

---

[36]Otherwise, the essential obligation of the BRC would be devoid of significance. Rightly, case law does not admit it.

[37]The *Biobank* network has already implemented a pricing service grid. However, it has no meaning and credibility unless the corresponding services have a certain level of quality.

In general, the convention should rigorously supervise the terms of partnership to ensure the best possible development from the collection. This rigorous management, which is the price of the contract effectiveness, will happen through provisions ensuring strict control of the respect of the purpose of the collection and the conditions of its operation. They may seem too drastic and, despite the singularity of the object of the convention, they should be called into question in the name of competition law,[38] the reason for that being that private law has its limitations and must give way to the special regime of administrative contracts.

References

Chassang, G., Chambon Thomsen, A., & Rial-Sebbag, E. (2016). *Ethique et réglementation des biobanques de recherche*. Paris: INSERM.

Chenut, C.-H. (2003). *Le contrat de consortium*. Paris: LGDJ.

Rochfeld, J. (2014). Penser autrement la propriété: la propriété s'oppose-t-elle aux "communs". *Revue internationale de droit économique, 28*(3), 351–369.

[38] Art. L. 442-6 of the Commercial Code.

Chapter 13
The Governance Framework of North American Biobanks: A Rapidly Evolving Policy Patchwork

Yann Joly and Miriam Pinkesz

13.1 Introduction

In recent years, a growing number of biobanks and associated research projects have emerged in North America. This paper intends to map out the governance framework of biobanks in Canada and the United Sates (US). The governance of biobanks can be characterized as a threefold enterprise: legal instruments, such as legislation and case law will apply to many biobank processes and activities; ethics codes and institutional policies provide a source of non-binding governance rules; and finally, biobanks themselves often develop their own governance policies. An overview of the current legal landscape reveals that Canada and the US have not adopted biobank-specific regulations. However, as biobank activities touches upon various areas of law, different laws and legal principles will collectively shape their governance framework. As such, a comprehensive presentation of applicable legal and ethics norms pertinent to biobank research requires that various biobanking practices first be identified. In doing so, correlating rules are more easily identifiable in the ensemble of applicable laws and ethics norms.

This paper focuses on the governance of data and biospecimens contained in biobanks, focusing on the following key legal and ethical issues: free and informed consent, fiduciary duties, privacy and property rights. The applicability of a given law or ethics policy to biobank research depends on various important factors. These include, among others, the nature of the materials used (i.e., biospecimens or data, or both), whether the material is de-identified and the professionals that are processing the data/samples or running the biobank (i.e., physicians, researchers, government employees). Legislations on these issues are broadly formulated, which often leaves gaps in their applicability to biobanks. Ethics and institutional policies and

Y. Joly (✉) · M. Pinkesz
Centre of Genomics and Policy, McGill University, Montreal, QC, Canada
e-mail: yann.joly@mail.mcgill.ca; miriam.pinkesz@mail.mcgill.ca

X. Bioy (ed.), *Public Regulation of Tumor Banks*,
https://doi.org/10.1007/978-3-319-90563-1_13

professional norms therefore attempt to fill these gaps through more precise guidelines. Importantly, biobanks are often partnerships with other institutions or government agencies, or are funded by the government. This may render biobanks' classification according to their source of funding (public or private, or provincial/state or federal challenging) misleading.

13.2 Canada

Biobanks in Canada are regulated through various legal domains including: medical law, criminal law, civil law, patent law and constitutional law. Due to Canada's unique bi-jurisdictional legal system and complex constitutional distribution of legislative power, in some instances, only federal or provincial laws will apply, or both.[1] As such, it can be said that "Canadian biobanks operate in a strange intersectional legislative space" (Saulnier and Joly 2016, p. 8). Hence, before ascertaining whether federal or provincial legislation applies to a biobank, various elements must be determined such as: whether the biobank is considered a federal or provincial entity or agency, whether the biobank is federally or provincially funded and whether the activities of the biobank engage federal or provincial legislative competences.

13.2.1 Legal Framework

13.2.1.1 Free and Informed Consent

Biobank research characteristically involves biospecimens and associated health data. Biological samples are often removed during clinical care, but can also be obtained solely for research purposes. Because biobanks collect and use bio-samples and data, they need to adhere to relevant consent requirements. Free and informed consent is a prerequisite for health-related procedures and research. The two contexts, however, must be differentiated as each requires a different standard of "informed consent". The Supreme Court of Canada established that there is a higher standard of disclosure of information in the context of 'pure research' than in the

[1]For example, both federal and provincial governments have jurisdiction over healthcare. Conversely, the federal government was granted exclusive jurisdiction over criminal law, which gives it authority over narcotics, devices and drugs, including blood and blood products, which can be relevant to biobanking. See *Constitution Act, 1867* (UK), 30 & 31 Vict, c 3; *Food and Drugs Act*, RSC 1985, c F-27.

clinical or therapeutic care context.[2] This stringency therefore requires research biobanks to provide more precise information to participants.[3]

The Supreme Court of Canada's stringent consent requirements are ill-suited to apply to research biobanks developed to conduct longitudinal studies since little information regarding future research, potential risks and future findings will be known at the time of consent. In addition, biobanks often share data and samples with other researchers and institutions, which further hinders biobanks from meeting a higher standard of informed consent disclosure.

13.2.1.2 Free and Informed Consent in Québec Civil Law

The Civil Code of Québec[4] (CCQ) contains specific provisions concerning the removal of biospecimens for fundamental and clinical research involving humans applicable to the biobanking context.[5] The CCQ requires that consent be obtained

---

[2]*Pridham v Nash Estate (1986) 57 OR2d 347* (HC) [*Pridham*]*; Brushett v Cowan* (1990), 69 DLR (4th) 743 (Nfld CA) [*Brushett*]*; Weiss v Solomon*, [1989] RJQ 731, 48 CCLT 280 (CS Qué) [*Weiss*].

[3]The Supreme Court of Canada requires that research participants be given all information that a reasonable person might be expected to consider before consenting to participate in the research. This means, for example, that a research biobank's consent forms would have to disclose rare risks, whereas biobanks solely involved in clinical care would not. Nevertheless, the scope of the disclosure and the kind of information that must be disclosed also depend on the facts of each case. In addition to the 'informed' component, consent must be voluntary. Jurisprudence warns that failure to obtain free and informed consent prior to a procedure that intrudes on a person's body can amount to the tort of battery or an action in negligence. *Halushka v University of Saskatchewan*, 1965 CanLII 439 (SK CA) [*Halushka*] at para 29; *Taylor v Hogan*, [1998] NJ No 14, 160 Nfld & PEIR 93 (Nfld. CA) [*Tayolr*]; *Hopp v Lepp*, [1980] SCJ No 57, 112 DLR (3d) 67 (Supreme Court of Canada) [*Hopp*]; *Reid v Maloney*, [2011] AJ No 1327, 2011 ABCA 355*; Malinowski v Schneider*, [2012] AJ No 759, 2012 ABCA 125[*Malinowski*]; *Reibl v Hughes*, [1980] 2 SCR 880 [*Reibl*]; *Starson v Swayze*, [2003] SCJ No 33, [2003] 1 SCR 722 (Supreme Court of Canada) [*Starson*]; *Gilgorevic v McMaster*, [2012] OJ No 721, 2012 ONCA 115 (Ont CA) [*Gilgorevic*].

The Supreme Court of Canada laid out the requirements for voluntary consent in the context of clinical care: consent is voluntary when it is free and without coercion, undue influence or misrepresentation. The application of this principle is visible in numerous biobank consent forms, which highlight the voluntary nature of the participation. For example, some biobank consent forms emphasize that the decision to forego participation will not affect clinical care. Furthermore, for consent to be valid, it must be given by a capable person. This means that the individual must understand the nature, purpose, risks and benefits of the proposed treatment and the consequences of one's decision to undergo or forego the procedure. See *Reibl, supra* note 4 in Chap. 2. *Mulloy v Hop Sang*, [1935] AJ No 8, [1935] 1 WWR 714 [*Mulloy*], *Murray v McMurchy*, [1949] BCJ No 74, [1949] 2 DLR 442[*Murray*], *Malette v Shulman* (Ont CA), 1990 CanLII 6868 (ON CA) [*Malette*]; *Edwardson v St Joseph's Healthcare Hamilton*10/26/12 [2012] OJ No 5032, 2012 [*Edwardson*].

[4]Civil Code of Québec, CQLR c CCQ-1991.

[5]The only other legislations that address the donation of tissue for research is *Human Tissue Gift Act*, CCSM c H180, s 9 and the *Human Tissue and Organ Donation Act*, SA 2006, c H-14 5, s 3. However, the legislations do not address the same issues as the CCQ.

prior to any care, including specimen taking and removal of tissue.[6] Aside from the requirement of consent for the alienation of biological material, the risks incurred to the participant must not be disproportionate to the benefit that may reasonably be anticipated (Kouri and Philips-Nootens 2012, pp. 516–517; Deleury and Goubau 2002, p. 145).[7] The CCQ requires the same condition for one to consent to research that could interfere with bodily integrity with the condition that the research project is approved and monitored by a research ethics committee.[8]

Article 22 of the CCQ requires that where biological material is removed as part of medical care, additional consent be obtained for that material to be used for research purposes.[9] Due to the specificity present in Québec civil law,[10] it is questionable if broad consent is permissible, because it allows original researchers as well as future researchers to use participants' biospecimens and data in a wide range of future research projects unknown at the time of recruitment (which may not meet the informed consent threshold currently set by case law) (Allen et al. 2013, p. 87).

13.2.1.3 Ethics Policies

The most widely used Canadian ethics policy, the TCPS2,[11] is a product of three federal research agencies,[12] to promote the ethical conduct of research involving

---

[6]Art. 11 CCQ.

[7]Scholars have pointed out that that the CCQ does not specify to whom the benefits and risks are directed. In the case of clinical and therapeutic care, the Québec legislature has indicated that the benefits and risks incurred apply to the patient. However, there has been no clarification for the research context. As such, the vague wording of the CCQ allows for the possibility that human-subject research benefit society or science rather than the participant herself. Nevertheless, scholars highlight that, according to the CCQ, research of a population group requires that the benefits to the group (i.e. persons of the same age category, or of the same illness, etc.) should outweigh the risks. The particular wording of the CCQ therefore requires population biobank researchers to ensure that the potential benefits outweigh the possible risks to the participants, whereas biobank researchers not dealing with population groups may have more leeway to seek benefits for society as a whole rather than the participants themselves. Nevertheless, these requirements necessitate that before consent to research is obtained, researchers evaluate the possible risks and benefits, and ensure that the benefits outweigh the risks (either to the participant, population group, or to society).

[8]Art. 20 CCQ.

[9]Art. 22 CCQ. As such, a cancer patient in the province, for example, must consent to the removal of a tumor in the context of medical care, and must provide additional consent for the removed tumor to be used in research as is often conducted by tumor banks.

[10]Such as ensuring that the risks incurred are outweighed by the benefits prior to patient consent.

[11]Tri-Council Policy Statement (2014).

[12]The Canadian Institutes of Health Research, the Natural Sciences and Engineering Research Council of Canada, and the Social Sciences and Humanities Research Council of Canada.

humans.[13] It has a specific chapter (12) on the use of human biological materials for research. According to the TCPS2, "consent" is defined as "free, informed and ongoing consent".[14] The TCPS2 makes a point to highlight that consent should be an ongoing process that is maintained throughout the research project.[15] The policy also lists all the information that is required for consent to be informed.[16] Certain elements will not be applicable to all research projects. In such cases, the researcher should explain to the research ethics board (REB) why the requirements should not apply to the particular research project.[17]

13.2.2 Privacy

As Canada does not have federal or provincial biobank legislation (Austin and Lemmens 2009, p. 115), it is challenging to identify a uniform legal standard for biobank privacy rules. Nevertheless, privacy protections provided by the Charter,[18] privacy legislation[19] and health information legislation[20] are applicable to biobanks more generally. It is important to note that in the context of biobanking, participants are owed privacy protections over both their personal health data and biological samples (Saulnier and Joly 2016, p. 8). Yet, Canadian legislation has not clearly distinguished between the two, and both are likely subsumed under "personal health information" (Saulnier and Joly 2016, p. 8).

Generally, biobanks must protect participants' reasonable expectation of privacy.[21] This expectation is important to delineate in the biobank context as it is different from other research or healthcare contexts. In biobank research, genetic data relays intimate health information, which may be widely shared or used for very

[13]Panel on Research Ethics (2017). Although the policy specifically applies to all institutions eligible to receive or administer funding from noted federal agencies and to members of those institutions, it has been extensively adopted by other agencies and institutions across Canada.

[14]TCPS2(2014), p. 25-4.

[15]*Ibid,* Art. 3.3.

[16]This includes, the purpose of the research, the identity of the researchers and funders, foreseeable risks and potential benefits, the expected duration of participation and an explanation of the responsibilities of the participant (TCPS2(2014) Art. 3.2.).

[17]*Ibid* at 28.

[18]The Supreme Court of Canada has interpreted Section 7, which guarantees the right to life, liberty, and the security of the person to include the protection of privacy in the context of medical records under "liberty". Section 8 of the Charter makes a more explicit mention of the right to privacy as it protects from unreasonable search and seizure. The Supreme Court of Canada has found that it is an invasion of personal privacy, under Section 8, when a person's body is used without consent to obtain biological information. See *R v O'Connor*, [1995] SCJ No 98, [1995] 4 SCR 411 (Supreme Court of Canada) [*O'Connor*]; R v Dyment [1988] 2 SCR 417[*Dyment*].

[19]*Privacy Act* RSC 1985, c P-2.

[20]*Personal Information Protection and Electronic Documents Act* (SC 2000, c 5) [PIPEDA].

[21]Panel on Research Ethics (2017).

long periods of time. As such, the expectation of privacy will need to consider the realities of biobank research.

13.2.3 Fiduciary Duties

13.2.3.1 Fiduciary Relationships and the Common Law Trust

It is widely accepted that physicians owe fiduciary duties to their patients. However, the law does not provide a definitive answer as to whether researchers owe the same duties to research participants. A determination that such a duty exists is founded on whether biobanks are trusts or whether a fiduciary relationship exists, and would mean that biobank researchers would have to treat participants, biospecimens and data with significantly more care.

In the common law, fiduciary obligations arose from the trust which is characterized as an arrangement by which property is managed by a trustee on behalf of a beneficiary.[22] As such, trustees have the obligation to act entirely in the interest of the beneficiary.[23] In a traditional trust model, researchers or the biobank act as trustees, the biospecimens and data constitute the trust property, and the research participant is the beneficiary. The critical distinction to be made in the case of research-oriented biobanks is that they do not conduct research for the sole benefit of the participant, but rather, for society at large (in a public infrastructure project) and the particular institution/company (in the case of a private biobanks) (Knoppers and Zawati 2011, p. 1182). As such, research biobanks diverge from the traditional trust model.

The idea that biobanks are a form of trust has been deliberated in literature over the years, and different trust models have been suggested, such as the widely-cited charitable trust model proposed by Winickoff (Winickoff and Winickoff 2003). Although biobanks do not fit the traditional common law archetype of trusts, biobank researchers are not necessarily precluded from owing fiduciary duties to participants. Outside the context of a legal trust, a party becomes a fiduciary where it undertakes to act for the benefit of another and has considerable discretionary power.[24] More precisely, fiduciary relationships characteristically involve an undertaking of loyalty vis-à-vis the beneficiary.[25]

---

[22]See, e.g. *Nolan v Kerry (Canada) Inc.*, [2009] SCJ No 39 at para 187 (Supreme Court of Canada) [*Nolan*].

[23]*Ibid.*

[24]*Hodgkinson v Simms*, [1994] SCJ No 84, [1994] 3 SCR 377 (Supreme Court of Canada) [*Hodgkinson*]; *Galambos v Perez*, [2009] SCJ No 48, [2009] 3 SCR 247 [*Galambos*].

[25]Canadian jurisprudence outlines indicia which indicate the existence of a fiduciary relationship outside the context of traditional fiduciary cases, such as lawyer-client relationships. The existence of such a relationship depends on the scope for the exercise of discretionary power, whether the power or discretion can be exercised unilaterally so as to affect the beneficiary's legal or practical

In the famous 1965 Saskatchewan Court of Appeals case, Halushka v University of Saskatchewan,[26] the court characterized the relationship between a medical researcher and a research participant as fiduciary.[27] The court determined this status based on the particular position of the parties: the research participant relied upon the special skill, knowledge and experience of the researchers. Similarly, biobank research involves a degree of reliance on the researchers by participants; participants rely on the validity of information given to them and they expect that their interests will be protected. However, biobank research can be distinguished from the type of research evoked in Halushka in an important way: in Halushka, the participant endured a physical intervention that resulted in a physical injury. Conversely, in the case of biobank research, participants donate biospecimens which are studied after only minimal physical intervention has occurred. Participants do not undergo further physical intervention by biobank researchers. Moreover, where data and biospecimens are shared between biobanks, investigators often do not meet with participants, and may likely be dealing with de-identified samples or data. As such, the researcher-participant relationship in biobanking is generally only bound by obligations established in law and applicable institutional policies, and not by fiduciary duties.

13.2.3.2 The Québec Fiducie

Québec fiduciary law has a different starting point than that of Canadian common law. The Québec fiducie is a civilian adaptation of the common law trust, and was adopted in the province in 1879 by the Acte concernant la fiducie (Cantin Cumyn 2013). Although much uncertainty remains regarding whether biobanks can constitute trusts, the Superior Court of Québec indicated that it is possible. In 2011, the Québec-based biobank, Génizon Biosciences Inc., went bankrupt, leaving behind 50,000 biospecimens. The Superior Court of Québec subsequently mandated Génome Quebec, a provincial economic development organization at its request, to take charge of the management of data and biospecimens in the possession of the bankrupt biobank, as a "fiduciaire" ("trustee") of Génizon Biosciences Inc.'s data and samples.[28] The terminology used points to the understanding that Génome Québec was in the position of a fiduciary in the context resembling a purpose

---

interests, and a peculiar vulnerability to the exercise of that discretion. Based on these indicia, physician-patient relationships have been recognized by the Supreme Court of Canada as characteristically fiduciary in nature. See *Hodgkinson, supra* note 5 in Chap. 3; *Norberg v Wynrib*, [1992] SCJ No 60, [1992] 2 SCR 226 (Supreme Court of Canada) [*Norberg*]; McInerney v MacDonald, [1992] SCJ No 57, [1992] 2 SCR 138 (Supreme Court of Canada) [*McInerney*].

[26] *Halushka, supra* note 4 in Chap. 2.

[27] *Ibid* at para 29.

[28] Genome Québec (2012).

trust,[29] or a social trust as termed in Québec civil law.[30] In light of this, the purpose of the trust was to ensure "that the integrity of the biobank was preserved and subsequent research activities will be conducted in accordance with the established governance framework",[31] as is characteristic of the CCQ's social trust's scientific purpose.[32] Although there is ambiguity as to what exactly was meant by the terms "trustee" and "fiduciary", the case of Génizon Biosciences Inc. points to a seeming readiness of Québec courts to recognize the existence of a trust in the biobank context in some circumstances, or at least the existence of fiduciary duties owed to biospecimens and data.

13.2.3.3 The Duty of Good Faith

As the existence of fiduciary duties in biobanking remains unsettled, biobank researchers may still owe stricter duties to participants under Quebec's civil law obligation of good faith.[33] Although this duty is not fiduciary per se, in certain circumstances it must be exercised in a manner like the common law's fiduciary duty. For example, at times the duty to act in good faith requires acting with honesty and loyalty, which is characteristic of fiduciary duties (Baudouin and Jobin 1998, p. 111). This duty normally applies in a contractual relationship, where one party is in a privileged position with regard to information, and where the other party is in a position of reliance on the disclosed information, like in consumer contracts (Baudouin and Jobin 1998, p. 264).[34]

It is possible that research participants are owed a duty of honesty and loyalty, as consent forms may constitute contracts of adhesion (Laurie and Postan 2013; Cabanac and Giroux 2007).[35] Where this would be the case, the contractual duties owed by the biobank researchers to participants will be stricter than other

---

[29]Note that this resembles Winickoff's charitable trust model, as a charitable trust is a form of purpose trust.

[30]Art. 1270 CCQ.

[31]Genome Québec (2012).

[32]Art. 1270 CCQ. A trust constituted for a scientific purpose.

[33]The CCQ provides that all civil rights must be exercised in good faith, and that no right may be exercised in bad faith. Art. 6-7 CCQ.

[34]See *Bank of Montreal v Bail Ltée*, [1992] 2 SCR 554[*BMO*].

[35]See *Grimes v Kennedy Krieger Institute, Inc*, 366 Md 29, 782 A 2d 807 (Md 2001) [*Grimes*].; *Yearworth v North Bristol NHS Trust* [2009] EWCA Civ 37, [2009] All ER (D) 33 (Feb) [*Yearworth*]. The contractual character of consent forms can be found in the wording of the CCQ's antecedent, the *Civil Code of Lower Canada*. The old code intended to situate written consent (in consent forms) in a contractual context, indicating that consent forms should follow the general contract law theory ("*suivant la théorie générale des contrats*"). Furthermore, it has been suggested that if consent forms indeed constitute contracts, they would be contracts of adhesion, as the consenting research participant lacks bargaining power. See *Civil Code of Lower Canada,* 29 Vict, ch 41, (1865) Art. 20.

contracts.[36] Regardless of the existence of a contractual relationship, biobanks are still obliged to abide by relevant laws and usage and not to cause bodily, moral[37] or material injury to participants.[38]

The difference between biobanks owing fiduciary-like duties of good faith and actual fiduciary duties in a trust context is important. As per the general good faith requirement, a biobank owes the duty of good faith to the research participant or to the contracting party, in the case of a contract. However, where a fiduciary duty exists, as seen in the case of Génome Quebec,[39] the biobank would likely be characterized as a purpose trust. This creates fiduciary duties owed to the specimens and data in the bank's possession (as opposed to the donors of the samples and data). As such, the biobank would be obliged to treat biospecimens and data with integrity and for the purposes collected (i.e., for the purpose of the trust). The duty of loyalty, in this latter case, is owed to the social trust itself rather than the donor.

13.2.4 Property Rights in Biospecimen

Important property-related questions may arise in the context of research biobanks. Two predominant questions relate to the ownership and IP rights over bio-samples and data stored in biobanks, and the possibility of patenting biomedical inventions.

13.2.4.1 Property Rights in Biospecimens in Jurisprudence

On two noted occasions, Canadian courts have acknowledged ownership interests in biological material, namely sperm straws.[40] In 2010, the Supreme Court of British Colombia found that one may claim property rights in biospecimens even if one is prohibited from selling them.[41] In 2014, the Superior Court of Ontario concluded that human tissue excised for diagnostic purposes is owned by the institution or hospital as a component of the medical record, pursuant to applicable law.[42] The

[36]This is due to the imbalance of expertise and information, and the degree of reliance, which is characteristic of contracts of adhesion.

[37]Moral injury includes psychological injury, such as psychological distress resulting from defamation or a breach of privacy.

[38]Art. 1457 CCQ.

[39]See Sect. 13.2.3.2.

[40]JCM v ANA, 2012 BCSC 584 (CanLII), <http://canlii.ca/t/fr3z5>[*JCM*], *Lam v University of British Columbia*, 2010 BCCA 325 (CanLII), <http://canlii.ca/t/2b93j>[*Lam*].

[41]*Lam, supra* note 22 in Chap. 3.

[42]*Piljak Estate v Abraham*, 2014 ONSC 2893 (CanLII), http://canlii.ca/t/g75bb[*Piljak*].

above cases suggest that biobank host institutions, such as universities or hospitals, can claim ownership over biospecimens and data in their possession.[43]

13.2.4.2 The CCQ Rejection of Property Rights in the Human Body

Québec civil law is clear that the human body and its parts cannot constitute property (Kouri and Philips-Nootens 2012, p. 49). However, the status of permanently removed biospecimens is a more complex question. Québec scholars Kouri and Philips-Nootens consider that permanently removed biospecimens that are subsequently donated may be the object of a possible ownership right.[44] They point out that the key public policy reason behind the non-ownership of human body parts, such as organs, is to prevent possible abuses.[45] However, the removal and subsequent donation of non-regenerative, potentially vital, organs is very different from the donation of sperm, blood, or other regenerative biological material to a biobank. The latter arguably pose fewer public policy and ethical concerns (i.e. only minimal physical risk associated with the donation and a low risk of privacy breaches), thus warranting a more permissive "property" approach (Deleury and Goubau 2002, pp. 140–141). As such, according to them, bio-samples stored in biobanks may justifiably be considered objects of property rights.

Because the law has left the question of whether donated biospecimens can constitute property unanswered, hospitals, universities and other moral persons, or biobank investigators may still be precluded from having a right of ownership in biospecimens as such. Nevertheless, if this is the case, they may still have the right to use (usus) the biospecimens and to derive benefit from them (fructus) (Kouri and Philips-Nootens 2012, pp. 49–50).[46] Importantly, the exclusion of having a right of ownership in biospecimens does not prevent researchers from collecting and using biological material, and it possibly also does not bar biobanks from acting as trustees in relation to the biospecimens and data.[47]

[43]It is important to note that because there is no particular law in Canada regarding ownership of biospecimens or data, the biobank, the institution owning it, or the principle investigator can claim ownership rights. Nevertheless, biobanks, universities and hospitals generally have specific ownership and IP policies determining this matter

[44]*Ibid*. A key justification for allowing property rights in permanently removed biospecimens is that they often serve the public good, and do not pose public policy concerns that other samples do, such as organs.

[45]Namely, pressuring persons in economically precarious situations to sell bodily material. *Ibid* p. 51.

[46]The permanent alienation of biological material renders it commercializable and seemingly subject to certain rights arising from occupation.

[47]See Sect. 13.2.3.2.

13.2.4.3 Patents

Research on biospecimens derived from biobanks could lead to genetic and biomedical innovations (new diagnostic tests, precision therapeutics, risk prediction algorithms, etc.). Biobanks often initiate research projects and provide biospecimens and data to other research initiatives as well. It is therefore important for biobanks to have an established IP and revenue sharing policy.[48]

In Canada, patents are primarily governed by the Patent Act (1985),[49] which requires that an inventive step be involved in conjunction with a natural phenomenon (i.e. DNA sequence) for it to be patentable. There are two key Canadian cases concerning biological patents: The Commissioner of Patents v The President and Fellows of Harvard College (Harvard College) and Monsanto Canada Inc v Schmeiser (Monsanto). Both cases leave ambiguities in the discussion on human gene patents, because both concerned non-human higher life forms, and the cases are somewhat dated. Nevertheless, they offer two generalizable principles of Canadian patent law with regard to gene patents: (1) the characteristics of a biological invention must be due to human intervention, and not naturally occurring phenomena[50] (2) if otherwise meeting general patentability criteria, genetically modified genes, cells and procedures used to carry out genetic modifications are patentable.[51]

The Harvard College case concerned a genetically–engineered cancer-prone mouse ("oncomouse"), developed for use in cancer research. In that case, the Supreme Court of Canada found that the mouse was not patentable because, among other reasons, as a higher life form, all characteristics of the mouse other than the transgene were present "completely independently of human intervention".[52]

---

[48]For example, research on tumor samples from the Ontario Tumour Bank led to the patenting of a method for identifying pharmaceutical compounds for the treatment of solid tumor cells (US 9002653 B2). Publicly funded biobanks will often require their users not to patent samples and primary data to ensure that the material remains widely available for other users. Policies on IP rights on inventions derived from the data/samples are more permissive and variable.

It is worth noting that in Canada, copyright is not applicable to the contents of biobanks, such as data and biospecimens, and therefore cannot afford IP protections to the content of biobanks, as it only extends to original literary, dramatic, musical and artistic works. However, the law of trade secrets is relevant to the biobanking context and may offer certain rights and protections outside of patent law, as it gives individuals a right of action against anyone who discloses the protected information. See *Copyright Act*, RSC 1985, c C-42; *Uniform Trade Secrets Act,* (1989), Art. 7. For more information see, Joly 2012.

[49]*Patent Act* (RSC, 1985, c P-4).

[50]*College v Canada (Commissioner of Patents),* [2002] 4 SCR 45, 2002 Supreme Court of Canada 76 [*College*].

[51]*Monsanto Canada Inc v Schmeiser*, [2004] 1 SCR 902, 2004 Supreme Court of Canada 34. [*Monsanto*].

[52]The mouse was therefore not an 'invention' as per sections 2 of the Patent Act (1985); *College, supra* note 30 in Chap. 3.

The 2004 Monsanto case followed the Harvard College judgment's reasoning to the extent that it found that genes and cells are patentable under certain conditions. Monsanto concerned a genetically modified herbicide-resistant canola seed. In this landmark case, Monsanto did not claim protection for the genetically modified seed or plant, but rather, for the genes and modified cells that make up the plant (Prudham 2007). Here, the Supreme Court of Canada held that genetically modified genes, cells, and procedures used to carry out the modifications are patentable.[53] More controversially, the Court also found that patents over modified genes and cells can provide a de facto monopoly on plants or animals made up of the modified genes/cells (Prudham 2007, pp. 409–411).

Methods of medical treatment are, a priori, not patentable in Canada, whereas use of a compound to treat or test for a disorder may be valid, as long as the claim does not limit the skill and judgment of a physician.[54] However, substances used in medical treatments and processes for creating such substances are patentable as long as they meet the requirements of patentability and can be differentiated from medical methods.[55]

13.3 The US

Similarly to Canada, the US has a constitutional division of powers. The Tenth Amendment grants states jurisdiction over powers not delegated to the federal government, or otherwise forbidden to them under the Constitution.[56] These powers

[53]*Monsanto, supra* note 31 in Chap. 3.

[54]*Tennessee Eastman Co v commissioner of patents* [1974] SCR 111 [*Tennessee Eastman*]; *Apotex v Wellcome* 2002 21 Cpr 4th 499 (Supreme Court of Canada) [*Apotex*]; *Novartis Pharmaceuticals Canada Inc v Cobalt Pharmaceuticals Company*, 2013 FC 985, 440 FTR 1[*Novartis*].

[55]*Tennessee Eastman, supra* note 36 in Chap. 3 at 118. However, uncertainty arose in 2015, when the Canadian Patent Office released administrative guidelines for patenting diagnostic methods. The guidelines seemingly invalidate gene-patent claims in diagnostic method patent applications. Although gene-related subject matter does not seem to be precluded from patentability in Canadian patent law, the new guidelines do not bode well for the future eligibility of biological patents, particularly gene patents. A recent patent case, *Children's Hospital of Eastern Ontario v Transgenomic Inc* (*Trangenomic*), could have shed light on the exact position of Canadian patent law concerning gene patents, but the case was settled extra-judicially in March 2016. In *Transgenomic*, the patent claims in question were methods for identifying genetic mutations associated with Long QT Syndrome. The *Children's Hospital of Eastern Ontario* challenged the validity of the gene patents in 2014, and by 2016, the action settled with Transgenomic agreeing to allow Canadian public-sector labs and hospitals to use their "inventions" on a not-for-profit basis. See *Children's Hospital of Eastern Ontario v Transgenomic, Inc et al.* (14 May 2015) T-2249-14; See Examination Practice Respecting Medical Diagnostic Methods—PN 2015-02. https://www.ic.gc.ca/eic/site/cipointernet...nsf/eng/wr03945.html. See settlement at http://www.cheo.on.ca/uploads/genetics/Gene%20patent/CHEO-Transgenomic-Settlement-Agreement-Signed-2016-03-08.PDF.

[56]US Const amend X.

are generally matters of local concern, such as establishing local governments and issuing licenses.[57] The federal and state governments also share certain powers, including making and enforcing laws. Therefore, the activities of biobanks fall under both federal and state legislation, and the laws often vary from one state to another.[58] Hence, we also observe the lack of uniformity of the governance framework pertaining to biobanks in the US.

13.3.1 The Common Rule[59]

The Federal Policy for the Protection of Human Subjects ('Common Rule')[60] and parallel FDA regulations are the primary US pieces of legislation that address

[57]Wex Legal Dictionary, https://www.law.cornell.edu/wex/federalism.

[58]For example, states pass their own health insurance regulations and privacy laws, in addition to the existing federal laws.

[59]It is important to note that to be subjected to the Common Rule, a biobank must be involved in "human subject research" that is funded or otherwise subject to regulation by a federal department or agency. This raises three key requirements: (1) the biobank must conduct research, (2) human subjects must be involved in the research, and (3) the research must be supported by a federal body. See 45 CFR §46.101 (2009).

As for the first requirement, a biobank must engage in a "systematic investigation, including research development, testing and evaluation, designed to develop or contribute to generalizable knowledge". As such, a biobank that merely stores and "banks" biospecimens would not be subject to the Common Rule. As for the second requirement, a human subject is defined as "a living individual about whom an investigator (...) conducting research obtains (1) data through intervention or interaction with the individual, or (2) identifiable private information". The term "intervention" is further defined as requiring interaction (i.e. communication or interpersonal contact) between the researcher and the research subject. This means that the Common Rule does not apply to biobank research where there is no communication or interaction with the sample donor, as is often the case with secondary uses or research on biospecimens from another biobank or institution. However, a biobank will still be required to abide by the Common Rule where there was no interaction if the information being used is identifiable and private. According to the statute, private information includes, among others, medical records, or any other information that the participant can reasonably expect will not be made public. Identifiable information is defined as information from which the identity of the subject is or can readily be ascertained by the researcher or linked with private information. Research involving only coded specimens or data, where the researcher cannot access the key to the code, is not considered "human subject" research. As such, biobanks and biobank researchers will need to use individually identifiable private information for the Common Rule to apply. See 45 CFR §46.102 (d) and _.102 (f) (2009).

Often enough, biobanks will fulfill these requirements, and the Common Rule will clearly apply. However, it is also common for biobanks to use de-identified biospecimens obtained from another institution, such as a hospital, or specimens that were originally collected for another research purpose (i.e., secondary uses). In such cases, the research will generally not fall under the scope of the Common Rule.

[60]45 CFR §46.116 (2009). The Common Rule was drafted by the Department of Health and Human Services (DHHS). As the National Institute of Health (NIH) is an agency of the DHHS, all NIH-funded research is subject to the Common Rule. Additionally, most academic and healthcare

research involving humans.[61] As such, both are often applicable in the biobank context. Recent amendments to the Common Rule resulted in the Final Rule, which was published in the Federal Register January 19, 2017. The Final Rule will take effect on January 19, 2018. The Final Rule significantly changes the Common Rule with regards to consent, the establishment of new exemption categories for research and new institutional review board (IRB) requirements which should simplify biomedical research regulatory and ethics processes.[62]

13.3.1.1 Biobanking in the US

In 2012, Henderson et al. conducted the first national survey of the then estimated 636 biobanks in the US (Henderson et al. 2013, p. 1). Of the biobanks that responded to the survey, 78% were affiliated with an academic institution, 27% belonged to hospitals and 15% were part of a research institute (Harrell and Rothstein 2016, p. 2). Furthermore, most biobanks had multiple affiliations (Harrell and Rothstein 2016, p. 2). The federal government was found to be the largest funding source for 36% of biobanks, and funded 57% of respondent biobanks in the 5 years preceding the study (Henderson et al. 2013, p. 11). The survey confirms the hybrid nature of many biobanks. The funding sources, activities and affiliations of biobanks are all relevant in the determination of the applicable legal regime and governance of US biobanks.

13.3.2 Legal Framework

13.3.2.1 Free and Informed Consent

Federal Legislative Framework

Human subject research, as conducted by biobanks, is primarily governed by the Common Rule and FDA Regulations.[63] Both legal instruments require that the informed consent process include a provision of information about the research,

institutions conducting human subject research adhere to the Common Rule regardless of their source of funding. The FDA has its own legislation that also addresses human subject research conducted by the FDA. See US Department of Health & Human Services 2016.

[61] *Ibid.*

[62] 82 Fed Reg 7149 (2017).

[63] 21 CFR §§50.1-.56 (1980).

including its purpose, reasonably foreseeable risks, reasonably expected benefits, etc.[64] The Common Rule further requires that consent be free and informed, and in a language that the subject can understand.[65]

The Final Rule modified the Common Rule's consent requirement concerning identifiable information used for secondary research. As of January 19, 2018, a new exemption category will be introduced, permitting identifiable data and biospecimens to be used for secondary research without consent, where one or more of the following conditions are met: the information or biospecimens are already publically available; the information is recorded in such a way that the identity of the participant cannot readily be ascertained and the investigator does not contact subjects or try to re-identify information; the secondary research is regulated under the Health Insurance Portability and Accountability Act (1996) (HIPAA) or is conducted by or on behalf of a federal entity and involves the use of federally generated non-research information that was and continues to be subject to federal privacy protections.[66] The exemption applies to the secondary uses of biospecimens. In the context of research biobanks, an example of this would be clinical pathology biospecimens, like tumors, used for purposes other than the original clinical diagnostic.[67] Furthermore, the Final Rule demands an additional step for researchers to prevail themselves of the exemption, requiring that an IRB review and approve the adequacy of privacy and confidentiality protections of identifiable private information for which consent has not been obtained.[68]

Another important amendment is the allowance of broad consent for prospective, unspecified research for storage, maintenance and secondary use of identifiable private information and biospecimens.[69] This permits researcher to use broad consent to collect identifiable private information instead of having to resort to using non-identifiable anonymized information, or needing an IRB to waive the requirement for specific informed consent for a particular study.[70]

13.3.2.2 Jurisprudence

Unlike the stricter Supreme Court of Canada informed consent requirements for scientific research, US Jurisprudence provides ostensibly less demanding

---

[64]45 CFR §46.116(2009); 21 C.F.R.§§50.1-.56 (1980).

[65]45 CFR §46.116 (2009).

[66]45 CFR §46.104(d)(4) (2009).

[67]The exemption expressly does not cover any primary collections of biospecimens or information.

[68]45 CFR §46.111(a)(7) (2009); See 82 Fed Reg 7263 (2017).

[69]82 Fed Reg 7200, §__.104(d)(7) and (8) (2017).

[70]*Ibid.*

requirements. According to the case law, it is unlikely that biobank researchers who engage in pure research are legally required to inform participants of all their possible motives and uses of biospecimens and data. Greenberg v Miami Children's Hospital Research Institute (MCHRI)[71] (Greenberg) is illustrative of the American approach to consent requirements for scientific research. In Greenberg, parents of children afflicted with Canavan disease provided tissue and confidential information for research on the disease and the development of a genetic screening test. After learning about MCHRI's patent on the Canavan genetic test, the parents sued the principal physician-researcher and the MCHRI, claiming that informed consent was not provided, as they were not informed of the defendants' intention to seek a patent on their research and to profit commercially from it. In its judgment, the court declined to find a duty to provide information about the researcher's economic interests within the informed consent procedure for scientific research.[72]

13.3.3 Privacy

13.3.3.1 Federal Privacy Laws

Privacy regulations applicable to biobanks are both federally and state legislated. Additionally, each biobank follows institutional regulations or specific policy norms. At the federal level, the key privacy instruments are the US Constitution,[73]

[71] *Greenberg v Miami Children's Hospital Research Institute (MCHRI),* 264 F Supp 2d 1064 (SD Fla 2003) [*Greenberg*].

[72] *Ibid.* The decision may be surprising given the well-established principles of *Moore v Regents of the University of California* (*Moore*), where the court did conclude that disclosure of a researcher's personal interests unrelated to the patient's health, including both research and economic, must be disclosed in the consent process (*Moore v Regents of the University of California,* 271 Cal Rptr (2d) 481(1990) [*Moore*]) In *Greenberg*, however, the court distinguished the case at issue from *Moore* in that the physician-researcher was solely involved in research and did not have a therapeutic relationship with the participants. As such, investigators involved in scientific research are likely not required to disclose as much information as researcher-physicians involved in clinical care are. *Greenberg, supra* note 9 in Chap. 6.

[73] Although there is no explicit constitutional right to privacy, the US Supreme Court has interpreted the Fourth, Fifth and Ninth Amendments to include privacy protections. As previously noted, the Constitutional protection of privacy only applies to government biobanks. *Griswold v Connecticut, 381 US 479 (1965) [Griswold].*

the soon-to-be amended Common Rule, the Privacy Act of 1974,[74] FDA regulations[75] and HIPAA.[76]

Common Rule/Final Rule

The Final Rule, amending the Common Rule, provides that IRB review should be conducted for the new exempt categories based on their risk profile. Under some exempt categories, such as: 'where identifiable private information and biospecimens are used without consent', IRB review is required to ensure that

[74]The *Privacy Act* applies to the collection, maintenance, use and dissemination of information about individuals stored in systems of records by federal agencies. Under the Act, federal agencies are generally prohibited from disclosing records unless consent has been obtained from the individual concerned. To be covered by the Act, federal databases (or biobanks) must retrieve information on particular individuals. Importantly, biospecimens are not included under the scope of the legislation, thus limiting the applicability of the Act to biobanks. Furthermore, only US citizens and permanent residents are covered, therefore only permitting them to sue for breaches under the Act. This last requirement has implications for non-US donors of data shared with US federal biobanks.

Under the Act, federal agencies are generally prohibited from disclosing records unless consent has been obtained from the individual concerned. Nevertheless, certain exceptions to this rule apply, which are particularly relevant in the context of research biobanks. One exception is the "statistical data exception", which permits disclosure to a recipient who will only use the information for statistical research or reporting, and where the record is transferred without individually identifiable data. As such, biobank de-identified data may be disclosed in certain circumstances. The "routine uses" exception may also apply to biobank research; where a record was collected for research, it may be considered a "routine use" as per the definition given by the Act. See, *Privacy Act of 1974*, 5 USC § 552a (1974); § __.111(a)(7); See also 45 CFR §46.104(d)(2)(iii) (2009); __ 5 USC § 552a (2);__ § 552a(b)(5); __ § 552a(b)(1), __ § 552a(a)(7); Department of Justice 2015.

[75]FDA regulations generally mirror the Common Rule. These regulations apply to research on investigational drugs or medical devices with human subjects. Unlike HIPAA or the *Privacy Act*, FDA regulations apply to research on biospecimens and not to health data. Therefore, biobanks are subject to FDA regulations if they are engaged in clinical research involving biospecimens. Legal protections for personal FDA records, like the Common Rule, address informed consent and IRB review requirements. Additionally, the regulations make an exception to the informed consent requirement for non-identifiable specimens. 21 CFR § 50.1 (1980); __ § 355; __ § 812.3(p). See Harrell and Rothstein (2016), p. 9.

[76]The regulation applies to health plans, healthcare clearinghouses and to healthcare providers who transmit "health information in electronic form in connection with transactions for which the Secretary of HHS has adopted standards under HIPAA". As such, HIPAA will cover biobanks that fall under any of these categories (see US Department of Health & Human Services (2013)). The rule also applies to business associates of said entities. The term "healthcare provider" is broadly construed, and includes physicians, hospitals, hospice programs, among others. A business associate is defined as someone who creates, receives, maintains, or transmits PHI for a function or activity regulated by the Act on behalf of the covered entity. See 45 CFR § 160.103 (2009); 42 USC § 1395x(r), (s), and (u) (1965).

there are adequate privacy safeguards.[77] IRB review is also required where consent was not sought, or where broad consent was obtained.[78]

HIPAA

HIPAA implements the Privacy Rule which, very generally, prohibits the disclosure of individually identifiable health information.[79] Biobanks need to be associated with an activity other than research, such as providing patient care and submitting claims electronically, or provide services to a covered entity to fall under the scope of the Act. As such, hospital-based biobanks are covered entities (Harrell and Rothstein 2016, p. 11). Additionally, a biobank can be a covered entity if healthcare providers officially work for the biobank and perform collection functions (Harrell and Rothstein 2016, p. 11). A covered biobank is prohibited from disclosing PHI that is individually identifiable,[80] unless the Act requires or permits it.[81] Conversely, the Privacy Rule does not restrict the use or disclosure of de-identified health information.[82]

13.3.4 Fiduciary Duties

In the US, as in Canada, uncertainty looms over the question of whether investigators owe fiduciary duties to research participants. According to US common law, a fiduciary relationship is characterized as one of agency, in which the agent acts on the principle's behalf.[83] The indicia of a fiduciary relationship are present in the medical and therapeutic contexts due to the medical license and the imbalance of physicians' knowledge and expertise vis-à-vis that of patients.[84] It is generally

[77]45 CFR § 46.111(a)(7) (2009).

[78]__§104(d)(2)(iii).

[79]45 CFR 164.502 (2013).

[80]PHI includes information such as demographic data, which relates to the individual's past, present or future physical or mental health condition; the provision of healthcare to the individual; or the past, present, or future payment of the provision of healthcare to the individual. Individually identifiable health information further includes personal identifiers such as one's name, address, social security number and information for which there is a reasonable basis to believe it can be used to identify the individual. See 45 CFR Parts 103, 160, 164.

[81]See generally 45 CFR Parts 160, 164.

[82]I.e., information that does not identify the individual or provide a reasonable basis to do so. 45 CFR §§ 164.502(d)(2), 164.514(a) and (b) (2013).

[83]Restatement (Third) of Agency § 1.01 (2006).

[84]*Witherell v Weimer* 396 NE (2d) 268 (Ill App Ct 1979) [*Witherell*].

believed that researchers are not fiduciaries.[85] Unlike physicians who are governed by fiduciary duties, researchers are governed by federal research regulations, institutional policies, and contracts—where applicable.[86]

Fiduciary duties of researchers were further discussed in Greenberg. Although fiduciary law is generally a matter of state-law, the court pointed out that to find a fiduciary relationship between a researcher and a research participant, confidence would have to be reposed by one party and trust accepted by the other.[87] The court noted that "There is no automatic fiduciary relationship that attaches when a researcher accepts medical donations and the acceptance of trust, the second constitutive element of finding a fiduciary duty, cannot be assumed once a donation is given".[88]

Although case law and commentaries suggest that researchers and researcher-physicians who are not engaged in therapeutic care generally do not owe fiduciary duties to participants, the courts have left open the possibility that where the indicia of a fiduciary relationship exist, researchers may be found to be fiduciaries. In the context of research biobanks, it is therefore possible that a fiduciary duty is owed to a participant where the indicia of a fiduciary relationship are found. However, this is more likely to occur in the case where a biobank's principal researcher is also a participant's physician, and if tissue is donated in the context of clinical care, similarly to the facts seen in Moore.[89]

13.3.5 Property Rights in Biospecimens

13.3.5.1 Property Rights

American jurisprudence is on the whole conclusive that once excised, biological material becomes the property of the research or medical institution that collected and has possession of it, such as a university or hospital.[90] Courts have characterized

---

[85]Departing from the common trend, the Maryland Court of Appeals found that researchers' obligations may extend beyond federal regulations because research can give rise to special relationships resembling fiduciary relationships. *Grimes, supra note 19 in Chap.* 3.

[86]*Whitlock v Duke Univ* 637 F Supp 1463 (MD NC 1986) [*Whitlock*]; *Daum v SpineCare Med Grp* 52 (2d) 260 (Cal App) (1997) [*Daum*]; *Vodopest v MacGregor*, 128 Wash (2d) 840 (1996) [*Vodopest*].

[87]*Greenberg, supra* note 9 in Chap. 6 at 1071.

[88]*Ibid.*

[89]In *Moore*, the patient's physician and hospital profited from research conducted on the patient's tissue (in the context of clinical care), by developing and patenting a cell line without disclosing this information to the patient prior to removal.

[90]As such, in biobank research, neither the principle investigator nor the research participants have property rights in donated biospecimens. However, the principle investigator may have IP rights in inventions derived from the institution's bio-samples. See *Washington University v Catalona* 437 F Supp (2d) 985 (2006) *[Washington]*; *Tilousi v Arizona State Univ Bd of Regents*, No CV2005, 2006

excised tissues as gifts or donations, thus eliminating all the donor's property rights in them.[91]

13.3.5.2 Patents

Gene patents often raise concerns relating to abusive uses (restrictive licensing practices, blocking patents, reach through rights, etc.) which may interfere with future research and prevent publically beneficial uses. This concern was a central issue in Greenberg, where the plaintiffs contended that the genetic information derived from their sample donations should have remained a public good rather than be used to promote the commercial interests of the principal researcher and hospital.[92] Members of Congress introduced a number of bills influenced by public concern on this issue (Cook-Deegan and Heaney 2010).[93] These bills and similar propositions were generally not adopted by the legislature, yet American courts have begun to limit the scope of patent-eligible genetic material, biological phenomena and genetic tests. As such, although the legislation remains the same, the Patent Act now applies to a narrower scope of inventions.

The US patent regime is regulated by the Patent Act[94] which provides that "[w]hoever invents or discovers any new and useful process, machine, manufacture, or composition of matter, or any new and useful improvement thereof, may obtain a patent therefor".[95] As per the Act's criteria, patents are divided into four categories: processes or methods, machines, articles of manufacture, and compositions of matter, or improvements thereof. To be patentable, an invention must satisfy basic patentability criteria (novelty, non-obviousness, utility and disclosure).[96]

WL 4642922: 1-25 (Ariz Super 2006) [*Tilousi*], *Havasupai Tribe v Arizona Bd of Regents*, 204 P 3d 1063 (Ariz Ct App 2008) [*Havasupai*].

[91] *Moore, supra* note 10 in Chap. 6; *Greenberg, supra* note 9 in Chap. 6.

[92] The plaintiffs also took issue with the defendants' restrictive (and costly) licensing of the genetic screening test.

[93] For example, in 2002, Representatives Lynn Rivers and David Weldon introduced *HR 3967* aimed at exempting research and genetic diagnostic use from patent infringement liability to allow the unrestricted use of such inventions. The proposed law would effectively exempt healthcare providers offering genetic tests and scientists undertaking non-commercial genetic research from being sued by holders of gene patents.

[94] 35 USC § 101(2012).

[95] *Ibid.*

[96] The first is that the invention or discovery must be novel. In addition, the invention must possess a utility, and the invention cannot be obvious to someone skilled in the art. The invention also needs to be described and explained in a manner that permits someone who is skilled in the art to make use of it. A further requirement is that the claims of the patent must be clear and specific. Finally, the entity must be patentable; laws of nature, physical phenomena and abstract ideas are not patentable. See, 35 US Code § 101 (2012).

The Bayh-Dole Act (1980)[97] offers institutions, like universities, the first option to acquire rights on inventions arising from federal funding (Cook-Deegan and Heaney 2010), yet the government retains the right to patent the invention if the institution does not do so. The purpose of the Act was to promote the utilization of inventions generated from federally-supported research through the patent system (Cook-Deegan and Heaney 2010). Because of the Act, a biobank host institution is encouraged to patent an invention when the research culminating in the invention was federally funded.[98]

13.3.5.3 Jurisprudence

It can be said that US case law on gene patents has shifted from a permissive approach to a more restrictive approach over the years, beginning from Diamond v Chakrabarty[99] (Chakrabarty) to Ariosa v Sequenom[100] (Ariosa). In the 1980 Chakrabarty case, the US Supreme Court determined that a microorganism constituted a "manufacture" or "composition of matter" within the meaning of section 1 of the Patent Act.[101] Although the Chakrabarty patent was not the first US patent on a living organism, this landmark case was followed by a statement by the Patent and Trademark Office (PTO), announcing that it henceforth considered non-naturally occurring non-human multicellular living organisms, including animals, to be patentable.[102] The Harvard oncomouse would be the first of many patented transgenic animals, among them, chickens,[103] cows,[104] rabbits,[105] and monkeys.[106]

More than two decades after Chakrabarty, the US Supreme Court began repositioning itself towards a more restrictive approach concerning gene patents. In the 2012 case of Mayo Collaborative Services v Prometheus Laboratories (Mayo), the Court concluded that dosage regulating processes, a category that covers some genetic tests, are not patentable.[107] In a similar vein, the Supreme Court in,

---

[97]35 USC § 202 (2012).

[98]The Act also sets requirements to ensure that inventions made under it are used "without unduly encumbering future research and discovery". The Act grants the NIH the authority to limit a grantee's right to elect title, or the NIH can elect title itself when it determines that the restriction or elimination of the right to retain title to the invention will better promote the policy and objective of the Act. 35 USC § 200 (2012).

[99]*Diamond v Chakrabarty,* 447 US 303 (1980) [*Chakrabarty*].

[100]*Ariosa v Sequenom,* 88 F3d 1371 (Fed Ci. 2015) [*Ariosa*].

[101]*Chakrabarty, supra* note 41 in Chap. 6.

[102]US Patent and Trademark Office Notice (1987).

[103]US Patent No. 5,656,479.

[104]US Patent No. 5,750,176.

[105]US Patent No. 5,675,063.

[106]US Patent No. 5,489,524.

[107]The claim concerned methods for determining optimal dosages of thiopurine drugs. According to the Court, medical diagnostic methods are laws of nature and require something that transforms the

Association for Molecular Pathology v Myriad Genetics Inc (Myriad),[108] concluded that naturally occurring genetic sequences (and segmented natural DNA) are not patentable, while synthetic complimentary DNA (cDNA) otherwise meeting patentability criteria are patentable (Cartwright-Smith 2014, p. 289).[109] Following Myriad, the US PTO released an interim guidance[110] on patent matter eligibility to shed light on the implications of the recent decisions by the Supreme Court and the US Court of Appeals for the Federal Circuit.[111] The guidelines apply the courts' restrictive trend, thus narrowing the scope of eligible gene patents at the application level.

In 2014, the US Court of Appeals, Federal Circuit revisited the remainder of Myriad's patents, subsequent to a claim of patent infringement,[112] assessing whether the claim regarding the BRCA1 gene was patentable. The court concluded that it was not,[113] arguing that synthetically replicated gene sequences[114] were not patent-eligible because they were structurally identical to the DNA strands found in nature.[115] As such, even cDNA must have a distinguishable element from natural DNA to be patentable.

---

nature of the object or process to be patentable. *Mayo Collaborative Services v Prometheus Laboratories* 132 S Ct 1289 (2012) [*Mayo*].

[108]*Association for Molecular Pathology v Myriad Genetics, Inc, et al*, 689 F (3d) 1303 133 S Ct. 2107 (2013) [*Myriad*].

[109]The key object of the patent claim was a patent on a genetic test to detect the BRCA1 and BRCA2 genes. The genetic test comprised of two steps which were analyzed by the Court: the first involved separating segments of DNA containing the sequences of nucleotides normally present in the BRCA1 and BRCA2 genes, and the second involved the creation of a copy of the original natural DNA sequence, only containing exons (i.e., cDNA). The Supreme Court invalidated some of Myriad's patents, thus allowing wider use of the genetic test and resulted in the availability of cheaper BRCA1/2 testing. This fact speaks directly to previous concerns regarding the patenting of biotechnological innovations: the wide use of patents of potentially life-saving tests and discoveries can be abused at the cost of the social good. The Supreme Court addressed this in its decision explaining that "[p]atent protection strikes a delicate balance between creating 'incentives that lead to creation, invention, and discovery' and impeding 'the flow of information that might permit, indeed spur, invention'". See *Myriad, supra note 50 in Chap.* 6.

[110]Note that the interim guidance is not law.

[111]See 79 Fed Reg 74618 (2014). The guidelines were updated to evaluate the patent eligibility of biological materials by assessing whether the product has "markedly different characteristics" than its naturally occurring counterpart. *Ibid,* at 74623–74624.

[112]*In re BRCA1-& BRCA2-Based Hereditary Cancer Test*, 774 F (3d) 755 (2014) [*In re BRCA1-& BRCA2*].

[113]*Ibid* at 1244.

[114]I.e. the primers.

[115]*In re BRCA1-& BRCA2, supra* note 54 in Chap. 6. The judges explained that it does not matter whether gene sequences are synthetically replicated. Rather, it is important that the gene sequences are distinctly different from those found in nature to be patentable. As such, the court interpreted the Supreme Court's ruling to mean that a DNA structure with a similar function to that found in nature must have a unique structure that is different from anything found in nature to be patent-eligible. The fact that the DNA has a different function or that it is synthetically replicated does not, in itself, grant patentability. This ruling invalidated several more of Myriad's patents.

The narrowing trend was also taken up by the Court of Appeals Federal Circuit in the 2015 Ariosa case.[116] Here, the court found that non-invasive prenatal genetic tests were not patentable, although Sequenom was the first to discover the testing method in question. The method of discovering cell-free foetal DNA (cffDNA) through a maternal blood test was judged a natural occurrence because the existence of cffDNA in maternal blood is a natural phenomenon.[117] Furthermore, according to the Court, there was no alteration to the genetic information or creation of any sort.

13.4 Conclusion

Biobanks have become key infrastructure resources for biomedical research. They contribute to ground-breaking discoveries and facilitate big data longitudinal and population research (Kinkorová 2015). Current uncertainties regarding key governance aspects of biobanks can be explained by the rapid expansion of biobank research and practices, and the relative absence of case law on many of the challenging questions they raise. Among the more interesting recent changes, we have identified a possible shift in Québec property and fiduciary law potentially warranting property rights in biobanks' biospecimens, the US Final Rule amendments to promote biobank research and emerging restrictions vis-à-vis genetic patents in US jurisprudence. These changes aim at facilitating the development and management of biobanks' assets and activities in a context conducive to data sharing and open science. For example, the Final Rule's new exemption categories allow for potentially greater biobank research opportunities as the previous consent requirements have been relaxed.

13.4.1 Lingering Questions

The governance of biobanks remains an area of law that lacks precision and uniformity. There is no particular biobanking legislation at the federal or provincial/state levels. This is not necessarily negative as it gives policymakers the necessary leeway to develop specific ethical guidance to rapidly address emerging issues in this research field. However, this reality means that biobanks must also conform to more general laws pertaining to their various activities, such as informed consent and privacy.

While recent changes in the US contribute to rendering the biobank framework more predictable in this country, important questions remain unanswered in Canada. These include uncertainties relating to: the legal status of biobanks, the status of

---

[116] *Ariosa, supra* note 42 in Chap. 6.

[117] *Ibid* at 1156.

permanently removed biospecimens in Quebec, the permissibility of various consent models, (e.g. broad consent), uncertainties regarding gene patents and remaining ambiguities relative to whether biobanks can constitute a trust and owe participants fiduciary duties. Some of these questions are addressed in Canadian literature, by ethics guidance or through institutional policies, yet the legislature and the courts.

References

Allen, C., Joly, Y., & Granado, P. (2013). Data sharing, biobanks and informed consent; A research paradox? *McGill Health Law Journal, 7*(1), 85–120.

Austin, L., & Lemmens, T. (2009). Privacy, consent, and governance. In K. Dierickx & P. Borry (Eds.), *New challenges for biobanks: Ethics, law and governance* (pp. 111–122). Intersentia.

Baudouin, J.-L., & Jobin, P.-G. (1998). *Les obligations*. Yvon Blais.

Cabanac, J., & Giroux, M. (2007). Le formulaire de consentement à la recherche: incompatibilité entre le droit des personnes et le droit des contrats. *Revue de droit de l'Université de Sherbrooke, 37*, 235–274.

Cantin Cumyn, M. (2013). Réflexions autour de la diversité des modes de réception ou d'adaptation du trust dans les pays de droit civil. *McGill Law Journal, 58*(4), 811–825.

Cartwright-Smith, L. (2014). Patenting genes: What does association for molecular pathology V. Myriad genetics mean for genetic testing and research? *Public Health Reports, 129*(3), 289–292.

Cook-Deegan, R., & Heaney, C. (2010). Patents in genomics and human genetics. *Annual Review of Genomics and Human Genetics, 11*, 383–425.

Deleury, E., & Goubau, D. (2002). *Le droit des personnes physiques*. Yvon Blais.

Harrell, H. L., & Rothstein, M. A. (2016). Biobanking research and privacy laws in the United States. *The Journal of Law, Medicine and Ethics, 44*(1), 106–127.

Henderson, G. E., Cadigan, R. J., Edwards, T. P., Conlon, I., Nelson, A. G., Evans, J. P., et al. (2013). Characterizing biobank organizations in the U.S.: results from a national survey. *Genome Medicine, 5*, 3. The pinpoint for 373 is page 7.

Kinkorová, J. (2015). Biobanks in the era of personalized medicine: Objectives, challenges, and innovation: overview. *The EPMA Journal, 7*(4), 1–12.

Knoppers, B., & Zawati, M. (2011). Population biobanks and access. In P. Zatti & S. Rodota (Eds.), *Il Governo del Corpo: Trattato di Biodiritto* (pp. 1181–1194). Giuffrè Editore.

Kouri, R., & Philips-Nootens, S. (2012). *L'intégrité de la personne et le consentement aux soins*. Yvon Blais.

Laurie, G., & Postan, E. (2013). Rhetoric or reality: What is the legal status of consent forms in medical research ? *Medical Law Review, 21*(3), 371–414.

Prudham, S. (2007). The fiction of autonomous invention: Accumulation by dispossession, commodification and life patents in Canada. *Antipode, 39*, 406–429.

Saulnier, K., & Joly, Y. (2016). Locating biobanks in the Canadian Privacy Maze. *The Journal of Law, Medicine and Ethics, 44*, 7–19.

Winickoff, D. E., & Winickoff, R. N. (2003). The charitable trust as a model for genomic biobanks. *The New England Journal of Medicine, 349*(12), 1180–1184.

Chapter 14
Biobank Expertise and the Research Unit-Biobank Relationship

Catherine Bobtcheff and Carole Haritchabalet

We have placed biobank's expertise at the center of its relationship with research units[1]; this expertise not only guides biobank's decisions on whether to supply samples (and therefore on whether the project goes ahead), it can also improve the project's chances of success. Hence, we postulate that biobank's expertise determines its ability to evaluate projects and its willingness to contribute to producing successful innovations. Experienced biobanks are generally more selective than less experienced ones, but they should also be better at providing good-quality samples, thus improving the likelihood a project will be a success.

Expertise is not the only important factor in ensuring research is successful; involvement by the biobank's staff can also contribute to a project's success. Such involvement creates a moral hazard problem[2] because contracts between biobanks and research units cannot include unverifiable variables. Although some of the biobank's actions can be evaluated, others are almost impossible to quantify. Hence, a biobank has a certain amount of leeway to decide the extent of its involvement in a project, and it is this leeway, which no jurisdiction would be able

---

Analysis based on a theoretical model developed by Bobtcheff and Haritchabalet (2017).

[1]Whether a biobank focuses on a small number of collections or offers a wide range of samples impacts its expertise: specialized biobanks have greater expertise than generalist biobanks.

[2]Moral hazard arises when one party is free to change its behavior to the detriment of another party after a transaction has taken place.

C. Bobtcheff (✉)
Toulouse School of Economics, CNRS, University of Toulouse Capitole, Toulouse, France
e-mail: catherine.bobtcheff@tse-fr.eu

C. Haritchabalet
University of Pau and Pays de l'Adour (CATT), Pau, France

Toulouse School of Economics, Toulouse, France
e-mail: Carole.Haritchabalet@tse-fr.eu

X. Bioy (ed.), *Public Regulation of Tumor Banks*,
https://doi.org/10.1007/978-3-319-90563-1_14

to verify, that creates moral hazard. We use the term "effort" to describe any non-verifiable and therefore non-contractual involvement by biobanks.

The following analysis examines the way in which interactions between biobank's expertise and the conditions of exchange (terms of the contract) affect the amount of effort a biobank is prepared to provide. We identify the economic inefficiencies in this relationship and propose recommendations in terms of public policy. Our analysis sheds light on issues related to biobank specialization.

14.1 Modeling

Whether a research project is successful and of value to society[3] depends on both exogenous and endogenous factors. Exogenous factors include the general economic situation, the demand for the innovation the research unit is trying to develop, the current legislation, and the project costs; endogenous factors include the biobank's expertise, the number of samples used in the project, and the effort provided by the biobank. The following paragraphs describe the interactions and assumptions underlying our model of the relationship between a biobank, a research unit, and the value of the research project.

14.1.1 The Research Unit

Our model applies to research units wishing to launch a research project requiring biological samples stored in a biobank and aimed at producing an innovative product or process. The research unit is ready to invest a fixed sum, I, in the project. Because the project requires biological samples, it cannot go ahead if the biobank refuses to make the samples available.

The research unit plays a limited role in our model, which assumes that the amount to be invested is fixed. Once the research unit has decided to launch a project, we assume that it must decide how many samples it needs. It must also decide which biobank to work with. We discuss the impact of biobank expertise on this decision at the end of the paper. Whether the research unit is public or private has no incidence on our model.

[3] The project's social value includes all the returns (financial, academic, knowledge creation, etc.) it generates.

14.1.2 The Biobank

We consider a biobank with a given level of expertise or experience, α, and that can supply the samples the project needs. The research unit is aware of the biobank's expertise, which is linked to its strategic positioning (specialization, quality). Biobank's operations involve fixed and variable costs, which may depend on its expertise.

In our model, the biobank must make two decisions. First, it must decide whether to accept the project. If it does, it then must decide whether to exert effort to improve the project's chances of success. Our analysis is initially centered on the biobank's level of expertise. The decision to specialize is discussed later.

Biobank's experience is a key factor when it comes to evaluating projects, as more experienced biobanks should be more able to assess the project's potential for success. Biobanks can accept only projects that are likely to provide a positive return for the biobank. Once the biobank accepts a project, it must decide how involved it wishes to become. Involvement can be defined as any investment in the project by the biobank (additional work on samples, scientific collaboration with the research unit, etc.), including any non-verifiable and non-contractual "effort" exerted to improve the project's chances of success. We assume that the biobank is free to choose whether it contributes to the effort. This effort introduces an element of moral hazard into the relationship, as the biobank will exert effort only if it believes that the resulting benefits will cover its costs.

14.1.3 The Project's Social Value

Our main assumptions regarding the relationship between the endogenous variables and the project's social value are:

- H1: More experienced biobanks are better at detecting projects that are likely to succeed.
- H2: Projects accepted by experienced biobanks are more likely to succeed than those accepted by less experienced biobanks.
- H3: The greater the biobank's effort, the higher the probability the project will succeed. Because effort has a cost, systematic effort is not necessarily desirable. We assume that only biobanks with a certain level of experience will exert effort.
- H4: There is a socially optimal number of samples, that is, a number of samples such that the project's social value is maximized.

A project's social value, *V*, is the resultant of all the benefits (expected revenues from innovation) and costs associated with conducting the project (investment by the research unit, biobank's fixed operating costs, cost of producing samples, cost of effort). Undertaking a project is socially optimal if its social value is positive, and it is socially optimal for the biobank to be involved in the project if this involvement increases the project's social value. Innovation is said to be insufficient when projects with positive social value are not undertaken and to be excessive when projects with negative social value are undertaken.

14.1.4 Sharing the Project's Value

The research project's social value, V, is shared between the biobank and the research unit. The value (profitability) for each actor is the difference between the benefits obtained from participating in the project and the private cost of this participation. However, a project will only generate income for its stakeholders if it is successful.[4] If the anticipated values of the project to the biobank and research unit are denoted by V_b and V_r, respectively, then $V = V_b + V_r$.

The project's value is shared according to a contract specifying the payments the research unit must make to the biobank.[5] Our model considers a contract composed of a fixed transfer fee, F, a tariff for each sample provided, t, and a royalty rate, r, specifying the share of income from the project allocated to the biobank. Any effort contributed by the biobank is not verifiable and cannot be included in the contract. Hence, in addition to the project's investment cost, the research unit must pay the biobank the agreed sum for each sample and the fixed transfer fee. Because these two payments are made whatever the project's outcome, they provide a guaranteed revenue for the biobank, which it uses to cover all or part of its costs. These costs include a fixed cost, a variable production cost, and the cost of any effort the biobank contributes.

A contract covering the research project will be acceptable to both partners only if it provides both with a positive anticipated value. Conversely, both V_b and V_r must be positive, so the overall social value will also be positive ($V = V_b + V_r$). Consequently, only socially desirable projects will be undertaken. The relationship between biobanks and research units does not lead to excessive innovation, but it may lead to insufficient innovation, as some projects with positive social value may not be undertaken.

Once the biobank has agreed to a project, it will exert effort only if doing so will increase the project's profitability compared with not exerting effort. This raises the issue of under- or over-provision of effort.

---

[4]The revenue of innovation, R, is positive if an innovation succeeds. R is zero if an innovation fails.

[5]We assume that all project's returns can be expressed in monetary terms.

14.2 Results

We now examine how the relationship between the biobank's expertise, its decision on whether to exert effort, and the terms of the contract impact the innovation's social value. We approach this question from the perspective of a regulator wishing to maximize a project's social value, in which case the actors' decisions must create the greatest possible value for the society. We begin by examining the different components of the contract, then we look at the most important aspects of the relationship between the biobank and the research unit.

14.2.1 The Sample Tariff

The first issue is the tariff, t, paid for the samples. This tariff is both a means of transferring income from the research unit to the biobank in compensation for the cost of producing the samples and a key factor in determining the number of samples the research unit obtains from the biobank.

There is a socially optimal number of samples at which a project's social value is maximized (assumption H4). This number is attained when the sample's expected marginal value (the marginal value of a sample for successful innovation) is equal to its marginal cost.

If the decision is left to the research unit, it will choose a number of samples such that its marginal cost, the tariff, t, is equal to its marginal benefit, that is, its share of the sample's expected marginal value. This share is dictated by the royalty rate, r, which determines the share $(1 - r)$ of the revenues from the innovation received by the research unit. Thus, the marginal benefit and cost of a sample for the research unit are different from a sample's social marginal benefit and cost; therefore, there is no reason for the number of samples requested by the research unit to coincide with the socially optimal number. Consequently, sample price and royalty rate are factors that can reduce an innovation's effectiveness by leading the research unit to choose a sub-optimal number of samples. The research unit's demand for samples decreases as the royalty rate, r, and the tariff, t, increase.

For the biobank, the sample's marginal cost is the marginal cost of production. The biobank is compensated for producing a sample by the tariff, t, and its share of the sample's expected marginal value, r, which is paid by the research unit. Hence, the biobank's decisions on how many samples to offer are based on the difference between the marginal cost of production and the tariff, t:

- If t is higher than the marginal cost of production, the number of samples increases as r increases;
- If t is lower than the marginal cost of production, the number of samples decreases as r increases.

The sharing of the sample's expected marginal benefit between the two actors leads them to select a socially sub-optimal number of samples. For example, if t is zero, all the variable costs of sample production are borne by the biobank. Consequently, the research unit will try to obtain too many samples, whereas the biobank will not want to provide enough samples. If t is greater than the marginal cost, the increased marginal benefit for the biobank would lead it to produce an excessive number of samples.

However, by sharing the marginal cost of production, c, along with the income from the innovation, it is possible to calculate a tariff that would encourage both actors to choose the socially optimal number of samples. Calculating the tariff according to the rule $t = c(1 - r) < c$ allows the socially optimal sample production to be achieved. It is therefore optimal for biobanks to bear part of the cost of producing the samples and to provide them at a lower price than the cost of production.

14.2.2 Participation, Contracts, and Expertise

A contract (t, r, F) must guarantee both actors positive profitability so they both agree to participate in the project. Let us suppose that the tariff for the samples is set according to the rule stated above, that the transfer fee, F, is fixed, and that the biobank's effort can be observed (and therefore included in the contract).

The *participation constraints* for each actor restrict the possible values of the royalty rate to an interval denoted $[r_{\min}, r_{\max}]$ if the biobank does not exert effort or to $[r^e_{\min}, r^e_{\max}]$ if the biobank provides an observable effort.[6] These intervals are interpreted as follows. All royalty rates in the interval are possible solutions in the negotiations between the biobank and the research unit. A royalty rate outside the interval will prevent the project going ahead because at least one of the actors decides not to participate. The lower bound of the interval is the minimum royalty rate that must be assigned to the biobank for it to agree to participate in the project. The upper bound of the interval is the maximum royalty rate the research unit will agree to pay the biobank. These royalty rates are defined for every possible value of the biobank's level of expertise. Hence, these rates can be plotted as a function of the level of expertise. The value of the project for the biobank is zero along the curve $r_{\min}$ (or $r^e_{\min}$, if it exerts effort), whereas it is strictly positive for any higher royalty rate. The value of the project for the research unit is zero along the curve $r_{\max}$ (or $r^e_{\max}$, if it exerts effort), but it is positive for any strictly lower royalty rate (Fig. 14.1).

The issue here is not to determine the royalty rate that will arise from a negotiation. The outcome of negotiations between biobanks and research units is determined by the alternatives available to the actors should the negotiation fail. Hence,

[6]These intervals are included in [0,1].

Fig. 14.1 Minimum and maximum royalty rates and expertise

the amount of competition between biobanks is fundamental to understanding the outcome of such a negotiation.

Bobtcheff and Haritchabalet's formal analysis (2017) can be used to establish the conditions that favor these negotiations, that is, to understand how this interval is shaped by biobank's characteristics and decisions. First, it should be noted that the interval exists only if the biobank is sufficiently experienced (experience greater than $\alpha_{\min}$ or $\alpha^{e}_{\min}$), and that it widens with the biobank's level of expertise. This result stems from the positive relationship between the probability the project will be a success and biobank expertise (hypothesis H2). Next, the interval depends on how much effort, if any, the biobank exerts. An inexperienced biobank may become involved in a project but it will not exert effort ($\alpha_{\min} < \alpha^{e}_{\min}$). Because effort is costly, the biobank will only exert effort if it receives a larger share of the project's value, hence, the minimum royalty rate is not $r_{\min}$ but $r^{e}_{\min}$. The research unit, on the other hand, reaps the benefits of the biobank's effort without bearing its cost. Consequently, the maximum royalty rate the research unit agrees to leave to the biobank increases with effort from $r_{\max}$ to $r^{e}_{\max}$. The values $\alpha_{\min}$ and $\alpha^{e}_{\min}$, which allow room for maneuver in the negotiations and therefore increase the likelihood the project will go ahead, are the values of α which guarantee that the social value of the project is positive. Hence, all socially desirable projects can be undertaken.

The above analysis of the relationship between the terms of the contract, the biobank expertise, and the biobank effort focused solely on the problem of whether the different actors will *participate* in the project. We now turn to the problem of the moral hazard caused by the non-observability of effort, in other words, how to *incentivize* effort.

14.2.3 Effort, Contracts, and Expertise

Given the conditions under which the biobank will accept the research unit's request, when the biobank's effort can be specified in the contract, the royalty rate and fixed transfer are equivalent instruments. However, when the effort cannot be specified in

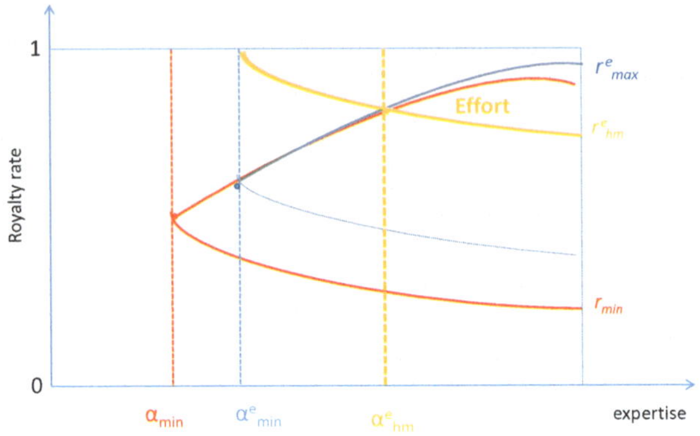

Fig. 14.2 Royalty rate, expertise and effort

the contract, the biobank will only exert effort if it serves its interests. Royalty rates and transfers are no longer equivalent: only a royalty rate that leaves all the value of the innovation to the biobank will encourage the socially optimal level of effort. Figure 14.2 shows (for a fixed transfer fee, F) the relationship between the biobank's expertise and the minimum royalty rate (r^e_{hm}) that solves the moral hazard problem, thereby making it possible to guarantee the biobank will exert effort.

The problem of moral hazard makes the exercise of effort less common, but allows the biobank to obtain a strictly positive value. Indeed, we assume that the royalty rate the research unit offers the biobank is r^e_{min}. By definition, the value for the biobank of participating in the project is zero if it exerts effort. Biobanks may accept to participate in such projects, but, because effort is not observable, they will prefer not to exert effort because, at this r^e_{min} rate, the value for the biobank of participating in a project without exerting effort is strictly positive. To encourage the biobank to exert effort, the research unit must offer a higher minimum royalty rate, r^e_{hm} ($r^e_{\text{hm}} > r^e_{\text{min}}$) such that the biobank **prefers** to exert effort. Conversely, the research unit must leave a strictly positive rent to the biobank, so, for a sufficiently experienced biobank, the value of a project is strictly positive.

Effort is socially efficient as soon as the biobank's expertise is greater than α^e_{min}. A biobank with this level of expertise will exert effort only if it receives all the income from the innovation ($r = 1$), in which case the research unit receives no income and therefore prefers the biobank not to exert effort. Effort is possible only when the biobank's expertise is greater than α^e_{hm} (where $\alpha^e_{\text{hm}} > \alpha^e_{\text{min}}$), whereas efficiency requires that all biobanks with expertise larger than α^e_{min} exert effort. In this case, the possibilities for exerting effort are restricted relative to the social optimum, and there is under-provision of effort for all levels of expertise $[\alpha^e_{\text{min}}, \alpha^e_{\text{hm}}]$.

14.2.4 *What Level of Expertise?*

The importance of expertise, demonstrated above, raises the question of the optimal level of expertise from both the research unit's and biobank's perspectives.

Research units consider two main considerations when selecting a biobank: the probability its project will be accepted and the value it expects to obtain from the project if it is accepted. Research units prefer the highest possible level of expertise, as the value they obtain from their projects increases with biobank expertise, even if the cost of sample production also increases.[7] However, because more experienced biobanks are more selective about the projects they participate in (hypothesis H1), the probability a project will be accepted decreases as biobank expertise increases. As Bobtcheff and Haritchabalet (2017) showed, the first effect always dominates the second, so research units always prefer more experienced biobanks.

As demonstrated in the preceding section, only projects carried out by the most experienced biobanks ($\alpha \geq \alpha^{e}{}_{\mathrm{hm}}$) guarantee a strictly positive profitability. In fact, this profitability can decrease as expertise decreases. This unintuitive result stems from the fact that an increase in expertise has several effects on the biobank's minimum profitability. More expertise increases the success of the innovation and therefore the biobank's profitability. But more expertise also allows a lower minimum royalty rate $r^{e}{}_{\mathrm{hm}}$, which reduces the biobank's profitability.[8] This second effect may outweigh the first, in which case the highest minimum profitability is obtained by the biobank with $\alpha^{e}{}_{\mathrm{hm}}$ expertise.

Although the minimum profitability decreases with increased expertise, greater expertise gives greater negotiating power, so more experienced biobanks are in a better position to negotiate favorable terms.

14.3 Economic Challenges of Networking

Expertise also plays a fundamental role in whether biobanks decide to participate in a network and share the fruits of their work with other biobanks. For example, what is the social value of additional expertise for a referring biobank charged with reviewing the research unit's request? Additional expertise can bring value to the project if it enables a referring biobank to realize it made a mistake when deciding whether to provide samples, or whether to contribute effort to the project. Therefore, a necessary condition for participation by a second biobank is that this participation increases the chances the project will be a success or increases the income from the project.

Clearly, inexperienced biobanks will benefit from joining a network, as being associated with more experienced biobanks gives them access to a larger number of

---

[7]See Bobtcheff and Haritchabalet (2017).

[8]As shown in Figs. 14.1 and 14.2.

projects and enables them to participate in more ambitious projects. In addition, working with specialized biobanks allows them to increase their expertise, which, in the long term, will have a socially positive effect on the quality of projects (Tykvová 2007). On the other hand, it is more difficult for an experienced biobank to associate with an inexperienced biobank as far as the inexperienced biobank's contribution to the project will have a lower social value than the experienced biobank's contribution (Casamatta and Haritchabalet 2007).

Moreover, this association aggravates the problem of moral hazard, as it is more difficult to get both biobanks to exert effort. The fact that each biobank benefits from the other's effort may lead one biobank to try and reduce its costs by exerting less effort. This free-riding behavior exacerbates the under-provision of effort. Because expertise is important to ensure effort, an inexperienced biobank is more likely to engage in such free-riding behavior at the expense of the more experienced biobank. Consequently, participating in a network is entirely beneficial for inexperienced biobanks, whereas experienced biobanks are likely to have to bear most of the costs (linked to information asymmetries) and will therefore be reluctant to join networks.

14.4 Public Policy Recommendations

The main features of the relationship between biobanks and research units are presented below, along with their implications.

(a) A simple pricing rule allows socially optimal sample production: the sample tariff must be lower than the marginal production cost. This tariff depends only on the biobank's marginal production costs; therefore, price discrimination according to whether the research unit is public or private is irrelevant.
(b) Experienced biobanks have greater negotiating power, but research units always prefer working with experienced biobanks.
(c) Only biobanks that are sufficiently experienced and that exert effort can achieve a strictly positive minimum profitability. Above a certain level of expertise, a costly strategy to increase expertise may not be appropriate, as the biobank will not get sufficient return on its investment.
(d) Networking is more easily accepted by generalist biobanks, which, almost by definition, have less expertise.

References

Bobtcheff, C., & Haritchabalet, C. (2017). Experience and screening in the management of innovation. Working Paper, 2017.

Casamatta, C., & Haritchabalet, C. (2007). Experience, screening and syndication in venture capital investments. *Journal of Financial Intermediation, 16*(3), 368–398.

Tykvová, T. (2007). Who chooses whom? Syndication, skills and reputation. *Review of Financial Economics, 16*(1), 5–28.

Chapter 15
Extra-Patrimoniality for the Patient

Sophie Paricard

Article 16-1 of the Civil Code provides, in paragraph 2, that the human body, its elements and its products cannot be subjected to property law. Further on, Article 16-5 provides that 'the conventions which confer a property value to the human body, its elements or its products are invalid'. Reading these texts, it seems that the valorization of tumor collections is difficult, but it is necessary to go beyond the obvious because the extra-patrimoniality, which should be defined, is actually only about the patient, leaving out of bounds the protective principles of the Civil Code in relation to all successive stakeholders of these biological samples, which might encourage the lawmakers to extend it.

But, even if one remains focused on the patient, what exactly does property law and property value mean? Why is there such a difference in vocabulary? Does property law equal property value? Are only free cessions allowed?

Parliamentarians are thought not to have shown legal rigor and seem to have confused preservation with venality. According to D. Fenouillet, "lawmakers do not seem to have used the notion of property law in its technical sense – an actual right or a claim included in heritage – but in a vulgar sense, a money-making sense" (Fenouillet n.d., *Jurisclasseur*). In that sense, F. Bellivier estimates that the provisions of the law of 29 July 1994 do not make the qualification of property impossible but simply ban the legal person to profit from the exploitation of the human body (Bellivier 1997, p. 184).

The main concern of the lawmakers, in the vote of the first bioethics law of 29 July 1994, has been, in effect, to prohibit the commodification of the human body at its source, that is to say at the time where the person would "dispossess themselves" (and the expression is not neutral) of an element or a product of their own body for cash. We all have in mind the character Fantine who, having sold her hair

S. Paricard (✉)
Institut National Universitaire Champollion, Albi, France

Institute in Private Law, University Toulouse 1 Capitole, Toulouse, France

X. Bioy (ed.), *Public Regulation of Tumor Banks*,
https://doi.org/10.1007/978-3-319-90563-1_15

for cash, did the same with her teeth. But with the advances of science, the possibilities of butchering an individual have truly materialized: one can give away their bodily products and especially their gametes (oocytes are especially coveted) and, in more serious cases, one can give away an organ which is part of a pair or which can regenerate (kidney, or lung, liver). The price of an organ can reach several thousands of euros on the black market and, officially, in California clinics, the oocyte of a sporty white girl who studies in a prestigious college, is worth a price.

That frightening bodily trade perspective, which means that a person may voluntarily give away for money an element or a product of their body, has therefore led to the drafting of these texts.

This confusion between ownership and venality comes from the fact that the discussions preceding the adoption of the law were not stringent concerning the legal terminology. The provisions of the bill gave rise to cryptic discussions, parliamentarians are not familiar with legal language, and, admittedly, they were not helped by the unbelievable wording of the original project which mixed up, among other things, extra-commerciality and payment.

Indeed, there have been many discussions about the legal status of the human body, its elements and its products. Principles such as the unavailability of the body have been used then abandoned demonstrating the difficulty of defining a legal status for the human body. The main confusion, as we will see, besides the occasional difference of language between public and private law specialists (Thouvenin, p. 36), comes from the fact that we have tried to use a vocabulary meant for things on the human body when its elements and products are related to a person and can only circulate in the context of a sample collection.

Therefore, our words are intended to clarify this vocabulary and to make articles in the Civil Code easier to read for our purpose—these articles are often greatly set aside compared to those of the public health code, which are more technical, more accurate and more efficient in a way—but also to highlight their character, somewhat offset in the light of the purpose of our conference, since the proposed protection of the patient is no longer effective. The possibility to bargain does not lie with the patient anymore but further in the chain, with the collector. Therefore, we shall not deal with data but focus on the actual tumor. The articles of the Civil Code should be revisited as such and so the key to this process is the extra-patrimoniality that we need to clarify with two certainties:

- a tumor is part of the human body
- a tumor is a harmful component of the human body

15.1 A Tumor Is Part of a Human Body

As a tumor is part of a human body, it is not a person's property (Sect. 15.1.1), and it can only be collected from a person's body (Sect. 15.1.2).

15.1.1 An Element of the Human Body Is Not Part of a Person's Property

In principle, a thing can go from one's property to another's. Indeed, from the point of view of the law, a thing can be owned, and so be part of a person's property. A thing is likely to go from property A to property B. However, there are things under a special status which cannot be the subject of such circulation, these things are said to be not for sale under former Article 1128 of the Civil Code; they are considered as unavailable. They are somehow frozen, the extra-merchantability referring to "the state of certain economic interests which cannot leave the property where they are and thus cannot change owner" (Paul 2002, p. 132).

Therefore, it has been really tempting to qualify these parts of the human body in the same way, so as not to include them on the traditional market of things and thus give them a certain aura of sacredness and, also, to consider them as unavailable. Jurisprudence prohibiting surrogacy agreements is thus based precisely on the principles of unavailability of the body and the state of persons, directly related to Article 1128. Under this article, the human body is not open to convention, and, therefore, it confirms the wrongfulness of conventions covering the person. For example, the following quote is an extract from a thesis precisely relative to the distinction between persons and things: "this text entails that proceeds of disposition relate only to the things that are for sale on the market. However, persons, by definition are not things for sale on the market or not for sale" (Andorno 1996).

This is how "the status of the human body has been entirely defined by reference to Article 1128 and the category of things that are not for sale" (Paul 2002, p. 190), which has become a kind of catch-all category created when it became a necessity to protect the human person, but "it is, in reality, out of ease of reference rather than out of accurate analysis that we reason in this way" (Terré and Fenouillet 1996, p. 17).

Extra marketability is only about circulation between properties. There are many conventions which are not about flow between properties, but about benefits involving the human body, and which are considered as lawful given that Article 1128 cannot validly oppose them: work contract, transfer contract (in particular, football players), prostitution. So, there has been a confusion which F. Paul perfectly explained in his thesis between extra marketability and extra-patrimoniality.

It has indeed been forgotten that "for a thing to be classified not for sale (...), it should exist as a thing" (Beignier 1995). Things that are not for sale are a subset of things (Paul 2002, p. 108). "The legal extra marketability does not prevent to be owner of a thing, but to negotiate this thing. It does not prevent property but venality" (Beignier 1994).

However, a body, its elements, and its products are not part of the property of a person: they cannot be owned, which prevents their circulation between properties. The human body, its elements and its products are extra patrimonial: they are not part of the patrimony of the person. This means that "the constituent elements of the sphere of the person cannot basically, substantially, be detached from the legal subject of which they are an integral part" (Paul 2002, p. 164) and as such cannot be part of the category of things. They constitute the category of person. Article 16-1 al. 3 is then fully coherent. Therefore, there is no possibility for bodily elements to circulate from a patrimony. They can come from a body.

15.1.2 A Bodily Element Is Part of a Person: It Can Only Be Collected from Their Body

Mr. Baud, in the case of the stolen hand, reiterates that "it is precisely since the human elements can be separated from the body, awaiting to be reincorporated that the case of a hand stolen from a transplant process is only jurisprudence fiction, not science fiction", that the lawyers' civil law conceptual system, which was created for things, has shown its limits (Baud 1993, p. 18). Lawyers have been challenged to fathom the body.

Taking a thing out of a patrimony is materially easy, common, a normal trade process, but collecting an element or a product from a body is essentially an attack on the physical integrity of a person, which, in principle, is a criminal offence. Therefore, it is essentially an abnormal act.[1] Collecting an element or a product from a person's body is an intrusive act as far as their physical integrity is concerned. This aspect seems to be forgotten when the collected element is said to be a donation, as if this common legal term included and contained that bodily sample.

However, it is not a donation in the common legal sense of the term as can be read in the Braibant report of the Council of State « Sciences de la vie. De l'éthique au droit »: "this is a donation (in the sense of blood donation), which is not a civil donation of a patrimonial nature" (Braibant 1988, p. 38). Indeed, there is no donation in the legal sense of the term: a donation implies the circulation of a thing from a patrimony (which is then reduced) to another.

Besides, the vocabulary used in bioethics is quite different from that used in relation to donations. We do not use the terms giver and donee but donor and recipient instead. A donation is never assimilated to a legal act that is likely to be cancelled based on common law (defect of consent). In short, "a bioethics donation is a false alienation. It is a donation only by name" (Prieur 1999). Bioethics donation is the purpose of collection but the collection process from the body of the person basically requires their consent.

Consent is not specifically about donation, but about physical integrity which requires the will to remove an element or a product from our body, and as there is no possibility to donate, because this element is not a thing, consequently, the person is not their owner. The only act a person can allow is the collection, the removal of a part of them which is considered as a violation of their physical integrity.

---

[1] This abnormal nature seems to prevent any qualification of thing by destination, beyond the fact that it is the person as such. An element, a product of the human body, is never intended to be separated from the person.

Article l. 1231-1 is indicative of this assimilation between collection and donation: “the removal of organs from a living person who donates them (...)”. The physical operation of the removal which has been legally agreed implies a bioethics donation. It is not distinguishable from it as the study of the revocation of consent shows.

Consent, previously expressed by a judge, does not constitute a legal act, even a unilateral one, and revocation does not break any legal act but only prohibits the violation of the physical integrity of the person. Consent is revocable without any formalities and at any time, which means up until the removal.

Also, the free nature of the process is linked to the act of removal as stated in Article 16-6: “no payment can be made to someone who consents (...) to the removal of parts of his body.”

And, the circle is complete, that is literal understanding of what Article 16-1 provides: “the human body, its elements, its products, cannot be the object of a property right”: the removal of an element from a body cannot be open to payment.

Beyond this characteristic shared by other bodily elements such as blood, sperm, or organs, a tumor has a special specificity: it is harmful to the human body.

15.2 A Tumor Is Different from Other Elements: It Is Harmful to the Person

Due to a tumor being harmful to a person, it cannot circulate, like other elements, from body to body (Sect. 15.2.1). Moreover, the removal of a tumor seems to be in the interest of the person and even medically necessary (Sect. 15.2.2).

15.2.1 No Circulation Is Possible from Body to Body

The circulation of human bodily elements usually occurs from one body to another: “the integrity of the human body can only be interfered with in the case of medical necessity for the person or in the therapeutic interest of a third party” under Article 16-3 al. 1 of the Civil Code. Although the expression ‘medical necessity’ when it comes to that person may include specific medical acts that have no therapeutic interest—such as an abortion—the purpose of bodily element removal to benefit a third party is clearly defined as having a therapeutic value which their body benefits from, so an incorporation in the literal sense of the removed element. This possible incorporation into someone else’s body, this vital primary value, makes it sometimes priceless.

Thus, the elements and products of the human body have somehow a primary, natural and sometimes even exclusive[2] purpose; that is to become part of a person's living body. They have a vital value whether they allow for life to continue (organs, blood) or give life (gametes). However, blood, tissues and cells can be removed with a scientific purpose and become part of a collection, not a body. They have a secondary purpose beyond their vital value.

As for the tumor, it is obviously never intended to become part of a person's body. It has no vital value, on the contrary. Therefore, the patient can never make a profit out of his tumor by proposing it to individuals. It is hospital waste. This was emphasized by "biobankers", who say that they even sometimes pick it "out of the trash": the tumor itself is worthless. What gives it value is, on the one hand, the data associated with it and, on the other hand, the expertise linked to it. It is given value only when it becomes part of a collection. Because of this, it would become part of a patrimony and would therefore even change qualification: elements and products of the human body become part of a patrimony.

15.2.2 A Necessary Removal

The will to oversee the person's consent in the context of a removal for donation purposes only makes sense because this collection does not have a medical necessity for the person. The violation of their physical integrity only rests on their will, and it became necessary to provide for rules to ensure the reality and the seriousness of their informed consent. Thus, consent to organ removal is very formalized, it includes a reflection period and an enhanced obligation of information.

However, the removal of a tumor or of some tumor cells always meets a medical necessity whether diagnostic or curative. Therefore, it is not necessary to give a framework to removal consent. Obviously, once the essential step of the removal has been taken, consent to the integration of this tumor in a collection is not very structured and can even be presumed, the person only having the possibility to oppose it.

The example of domino transplant allows us to better understand this reasoning: Organ donation from a living donor is highly regulated: it is forbidden from a minor or a protected adult[3] and consent for donation is, by law, a precisely detailed pathway. But as soon as the removal of the organ is a medical necessity, which is the case in domino transplant, consent to donate is presumed.[4] Donation is even

[2]This is the case for organs from a living person (excluding domino transplant).

[3]See Articles l. 1231-2 and l. 1241-2 of the Code of public health. Such a ban is even subject to penal sanctions (7 years imprisonment and 1 million euro fine, Article 511-3 para. 2 ASC).

[4]Art. L. 1235-2 of the code of public health: "the organs removed during a surgical procedure, performed in the interest of the operated person, may be used for therapeutic or scientific purposes, unless opposition is expressed by that person after they have been informed of the purpose of this use."

possible for a minor or a vulnerable adult. Donation is somehow separated from the removal and it is presumed as far as removal meets a medical necessity. Therefore, consent to the removal is the issue causing difficulty, and when the removal meets a medical necessity, consent to donation is less important since it is presumed.

This reasoning is also consistent in terms of the principle of secondary use in bioethics law foreseen by Article l. 1211-2 of the Code of public health correlated to a presumption of consent. It is possible to use elements and products of the human body for medical or scientific purposes other than that for which they were removed, unless the person who has been operated on expressly opposes it.

References

Andorno, R. (1996). *La distinction juridique entre les personnes et les choses à l'épreuve des procréations artificielles*, LGDJ, Bibliothèque de droit privé, T. 263, pref. F. Chabas.

Baud, J.-P. (1993). *L'affaire de la main volée*. Seuil: *Une histoire juridique du corps*.

Beignier, B. (1994). D. 1994, 163, note under Limoges, 10 May 1993.

Beignier, B. (1995). D. 1995, 561, note under Civ. 1st, 7 June 1995.

Bellivier, F. (1997). *Le patrimoine génétique humain, Etude juridique*, PhD Paris I, nb 215.

Braibant, G. (dir.). (1988). *Rapport du Conseil d'Etat, Sciences de la vie, De l'éthique au droit, Etude du Conseil d'Etat* (p. 38). La documentation française.

Fenouillet, D. Juris Cl. Civil, Art. 16 to 16-12, Respect et protection du corps humain. Protection de la personne, le corps humain.

Paul, F. (2002). Les choses qui sont dans le commerce au sens de l'article 1128 du code civil, pref. J. Ghestin, LGDJ, Bibliothèque de droit privé, T. 377, 2002, no. 164.

Prieur, S. (1999). *La disposition par l'individu de son corps*, PhD Dijon, 1998, Les Etudes hospitalières, coll. Thèses, T. 3, no. 228.

Terré, F., & Fenouillet, D. (1996). Civil law, *Les Personnes, la famille, les incapacités*. Dalloz, no. 17.

Thouvenin, D. La disponibilité du corps humain : corps sujet ou corps objet ?, Acts no. 49–50, Le corps aux mains du droit.

Chapter 16
The Moment of Patrimonialization

Marie-Xavière Catto

16.1 Introduction

Storing biological samples and questioning the rights various actors have on them necessitates the review of their conditions of removal as detached elements. These samples or collections of elements entirely determine any research in this field. These samples must be obtained. To the bank or the researcher they may only appear as previously collected samples, but they are the result of preliminary processing enabling their classification as detached elements. According to the elements and how they where removed — was there any infringement of the individual's physical integrity? was it necessary to inform the patient?— the rights of the persons and the banks differ. The protection of the provider depends both on rules directly applicable to them and on the organization of resources: if their use is optimal and centralized, the samples are well-preserved, and the research is the sole purpose for its access, fewer samples will be needed.

The persons who preserve, on the other hand, are the banks, i.e. "every entity [. . .] ensur[ing] the preservation and preparation of tissues and cells from a human body for scientific purposes as well as the preservation and preparation of organs, blood, its components and its derivatives," either for the needs of its own research programs (declarative regime, Article L. 1243-3 of the French Public Health Code), or to "cede" elements (authorization regime, Article L. 1243-4 of the French Public Health Code). Therefore, banks gather elements and these might, but not necessarily, be incorporated in collections. When the banks gather other elements, the Order of August 16th, 2007 specifies that "if the declared activity does not include the establishment or the use of biological sample collections and only focuses on irreversibly anonymized samples, the competent entity might only indicate the

M.-X. Catto (✉)
University Paris I Panthéon-Sorbonne, Paris, France

X. Bioy (ed.), *Public Regulation of Tumor Banks*,
https://doi.org/10.1007/978-3-319-90563-1_16

general theme of the research".[1] Collections, for their part, "refer to the collection of biological samples taken from a group of persons identified and selected according to the clinical and biological characteristics of one or several members of the group, as well as derivatives of these samples to be used for scientific purposes" (Article L. 1243-3 of the French Public Health Code). Thus, establishing a collection requires internal and scientific coherence for preservation: the collection must meet the needs of the "collector-researcher" (Le Roux et al. 2003) who compiles it. The legislative power did not consider the possibility for the banks to establish a collection in order to transfer elements (e.g. the Biological Resource Centres, BRCs). Nevertheless, the 2007 Order contains such possibility.[2] Although this capacity appears coherent, considering that their mission is to gather samples to ensure their availability, establishing a collection might also injure the physical integrity of subjects (and not only gathering previously removed elements). It is therefore difficult to understand that BRCs, which do not store samples for their own research programs, can be the initiators and not only the organization and preservation sites for previously collected samples.

Therefore, the legislative power organizes two types of banks: those which house samples essentially for themselves and those which provide access to samples. In both cases, the banks must be stocked with these biological materials. The first part of this review focuses on the modalities governing how banks or collectors request resources for scientific research and the conditions of access to different types of elements and products of the human body (Sect. 16.2). The "moment of patrimonialization," i.e. the moment when the person who provides a sample can no longer claim any right on said sample, depends on the collection process of the element swabbed or collected from the deceased or living person. Once these elements have been detached and preserved, the second part of this review analyzes the rights and duties of banks in respect to their use for scientific purposes (Sect. 16.3).

16.2 The Sources of Supply

The 1988 Huriet Law focused exclusively on acts containing regulations on research on human persons and not on elements, whereas the 1994 Law authorized removing samples for scientific purposes on living persons without specifying its legal application.[3] No specific regime therefore regulates the infringements on persons in the context of research on biological samples. The Law enacted on May 28th, 1996

[1]Order of August 16th, 2007 instituting a standard file for the statements and certification procedures necessary for the conservation and preparation of human body samples for scientific use, Official Journal of 08/18/2007, p. 13830.

[2]Order of August 16th, 2007 *above-mentioned.*

[3]Art. L. 672-4 incorporated in the Law No. 94-653.

introduced a provision relative to biological sample collections and, for the first time, enabled public and private law structures to collect samples *in order to establish collections* for genetic research purposes. However, due to the lack of text specifying its scope, this provision has never been applied (Rial-Sebbag 2009). Since these acts might affect physical integrity and the lack of precision in an alternative regime, doctors and researchers applied the amended 1988 regime (Le Roux et al. 2003, p. 807). On their side, "studies on [. . .] samples obtained without affecting physical integrity, or whose sampling constitutes an integral part of a necessary treatment"[4] were excluded from the scope of the 1988 Law at first. In 2002, IGAS (the French General Inspection of Social Affairs) observed that the vast majority of samples were collected at the end of diagnosis or in the context of routine care, and that as a consequence, swabbing elements was "very rare." (Dufourcq et al. 2002) When there is no infringement of the subjects' physical integrity during the sampling, several legal regimes coexist. Excluding specific provisions related to the importation of elements, the main resource of banks is the human body without legal personality, i.e. either the entire body which is not considered a person in the law (before birth or after death), or whose elements are already detached from the body (A). Conversely, if research involves the human subject, the applicable legislation is the 2012 Jardé Law (B).

16.2.1 The Main Supply: Bodies Without Legal Personality

The banks' supply originating from collected or swabbed elements from bodies without legal personality is governed by a triple regime that is increasingly regulated: (1) the general biomedical residues regime which simplifies access, (2) the even more regulated regime which applies to sampling from a corpse, and (3) the most regulated regime which applies to research on embryos.

16.2.1.1 Surgical Residues: A Simple Absence of Opposition

Definition The French Public Health Code defines residues of healthcare activity as "residues generated by the various activities of diagnosis, monitoring, preventive, curative and palliative that are part of human and veterinary medicine" (Article R. 1335-1 of the French Public Health Code). Residues are not classified by their nature, but rather by the process of their removal: they either manifest as the result of the classical therapeutic activity (diagnosis, prevention, treatment) or of aesthetic

---

[4]Protection des personnes dans la recherche médicale. Guide des textes législatifs et réglementaires, *BOSP*, numéro spécial, sept. 1991, t. II, p. 7.

surgery activity (Article R. 6322-17 of the French Public Health Code). This category includes samples collected in the treatment of the person, of a third party (the therapeutic aim being unsuccessful)[5]; samples from corpses collected during a post-mortem examination; or samples collected for scientific use, but preserved after the research that had justified the sampling, for another purpose. Specialists therefore provide residues from their own activities: amputation reveals bones collected by orthopedists, skin collected by doctors in burn units (IGAS 1993a), saphenous veins by the plastic surgeon after a stripping (IGAS 1993b, p. 8) blood sampled for analysis purposes by analytical laboratories (Harichaux 1988, p. 137), etc. These collected residues contribute to research on all elements and products from the human body.

The General Regime Governing Residues: Absence of Opposition Regarding residues, absence of opposition from the person is the main principle, as consent is not required. Article L. 1211-2 of the French Public Health Code provides that:

> The use of elements and products from the human body for medical or scientific use differing from the one for which they were removed or collected is possible, except if expressed opposition from the person, subject to this sample removal or collection, was duly notified of this other purpose beforehand.

The formula implying that potential use for other purposes "except in case of opposition by the person, after being informed of the object of this use," is replicated regarding "embryonic or fetal tissues or cells" (Article L. 1241-5 of the French Public Health Code), organs (Article L. 1235-2 of the French Public Health Code), and tissues, cells and products from the human body (Article L. 1245-2 of the French Public Health Code). Nevertheless, providing information before people may exercise their right of opposition is not always mandatory; two dispensations are provided in Article L. 1211-2 of the French Public Health Code:

> when the impossibility to find the concerned person is encountered, or when a Committee for the Protection of Persons, referred to in article L. 1123-1, who consulted with the person responsible for conducting the research, does not deem this information necessary.

The possibility to derogate from the information requirement is confirmed by Article R. 1123-22: "in case of use of elements and products from the human body for scientific uses with a change of purpose as compared to the consent initially given, the Committee, *if necessary [i.e. if it believes this information is necessary]*, ensures that the sponsor verified the absence of opposition from the persons." The article therefore specifies that if consent or absence of opposition are not always required, conversely, the Committee's approval is still mandatory. This general regime is supplemented by specific provisions for certain elements.

---

[5]Organs (Article L. 1235-2 of the French Public Health Code), tissues and cells (Article L. 1245-2 of the French Public Health Code), expired blood products and others (Article R. 1335-1 of the French Public Health Code).

Two Dispensations in the Regime of Absence of Opposition: Explicit Consent for Collected Cells from Placenta and from Embryonic and Fetal Tissues *Cells collected from the placenta*. The simple absence of opposition prescribed by the general regime is also used for the placenta. However, cells collected from the umbilical cord and placenta are exempt from this regime as consent is explicitly required.[6]

Embryonic or fetal tissues. As soon as 1998, the European Group on Ethics in Science and New Technologies (EGE) had formed an opinion seeking to derogate embryonic and fetal samples from the residues regime requiring "specific, free, informed consent from the woman or eventually the couple."[7] The 1999 scandals in England, much like those in France in 2005, concerning the anatomical parts of fetuses and stillborn children preserved in hospital freezers (Memmi 2011, p. 11), unknown to biological parents, sparked strong emotions. Obtaining the written consent of the subject, after "an appropriate notification on the purpose of such a sample," (Article L. 1241-5 and R. 1241-22 French Health Public Code)[8] for embryonic and fetal tissues and cells has been required since 2004. Additionally, the second exemption from the general residues regime requires that, to be removed, all samples must be included in a research protocol addressed to the French Biomedicine Agency, such as for samples on corpses.

16.2.1.2 Samples from Deceased Persons

Access Requirements: Information and Consent In 2016, the legislative power designated the computerized national registry as the principal mode of expression for the refusal of organ or tissue removal,[9] both in the therapeutic and scientific contexts.[10] Doctors are no longer required to consult the family. A relative (or the coordination team)[11] may still assert the refusal expressed verbally by the deceased but must mention in writing "the precise context and circumstances of this refusal."[12] In 2016, Article L. 1232-1 of the French Health Public Code expanded the information requirement about the therapeutic or scientific purpose (introduced in 2012) to include the nature of the part of human body removed.

[6]The dispensation incorporated in 2004 in Article L. 1245-2 of the French Public Health Code for the "cells of the umbilical cord blood and placental blood as well as cells from the umbilical cord and from placenta" does not apply to the latter which remains a residue.

[7]GEE, legal notice No. 11 of July 21st, 1998. Aspects éthiques des banques de tissus humains, p. 8.

[8]Decree No. 2007-1220 of August 10th, 2007, Official Journal of 08/14/2007, p. 13591.

[9]Law No. 2016-41 of January 26th, 2016, Art. 192.

[10]The French Biomedicine Agency noticed, between 1998 and 2007, seven refusals of samples for scientific purposes during the inquiries of the national file, Bilan d'application de la loi de bioéthique, Oct. 2008, p. 5.

[11]Supplement the Order of August 16th, 2016, Official Journal of 08/25/2016, point IIB.

[12]Art. R1232-4-4-III as modified by the Decree No. 2016-1118 of August 11th, 2016, Art. 2.

The Regime of Samples on Deceased Persons for Scientific Purposes In 1994, the legislative power favored organizations sampling for scientific purposes with a dispensation of authorization requirement. Organizations, regardless of their nature (public healthcare institutions, commercial societies, etc.) can, via a doctor from the bank or from a healthcare institution,[13] remove samples from lifeless, brain-dead, or non-beating-heart bodies. However, the 2004 Law[14] requires their compliance with a research protocol, addressed to the French Biomedicine Agency before its implementation (Article L. 1232-3 of the French Public Health Code for organs, Article L. 1241-6 of the French Public Health Code for tissues and cells and Article L. 1241-5 of the French Public Health Code for embryonic and fetal samples). The Agency provides a notification to the expeditor and addresses it simultaneously to the French Minister of Higher Education, Research and Innovation (Articles R. 1232-15 and R. 1232-16 of the French Public Health Code) who can disapprove it. The regime is declarative: the implementation of the protocol can start 2 months after the acknowledgement of receipt, except in cases of opposition (Article R. 1232-18 of the French Public Health Code). Conversely to rules applying to biomedical research (involving the human person), the French Health Products Safety Agency (AFSSAPS) is not competent and the Committee for the Protection of Persons (CPP) does not allow the research, but it only provides an opinion. The scientific data are succinct: the file allows a maximum of 4000 characters for the description of the research program (or programs), specifying that the aim is not necessarily to establish a collection (i.e. an internal coherence of research, which can only study anonymized elements).[15]

The 2006 Decree specifies that *the bodies of children declared as stillborn at the Civil Registry* can be returned to the family immediately, except in case of sampling within the 4 weeks following childbirth.[16] The necessity of such sampling must be justified (Articles R. 1232-17 and R. 1241-21 of the French Public Health Code).

The administrative formalities could have been simple if time constraints before placing a body into a coffin did not render this practical respect very difficult. Consequently, the use of this resource remains relatively insignificant in the establishment of collections (Hauw and Ardaillou 2009, p. 734).

16.2.1.3 Embryos

The Three Sources of Supply Embryos likely to be subjects of research are always conceived in vitro in the context of a medically assisted reproduction, responding to

[13]Order of August 30th, 1999 instituting a standard file; Official Journal of 09/01/1999, p. 13047.

[14]Law no. 2004-800 of August 6th, 2004—Art. 9.

[15]Order of August 16th, 2007 *above-mentioned*.

[16]Decree no. 2006-965 of August 1st, 2006.

the end of the parental project.[17] Legal texts regulate this eventuality in three situations (Mesnil 2017, pp. 9–10.) that represent the three sources of supply for the banks: in the context of a pre-implementation genetic diagnosis, if the couple decide to cease their project in case of a diagnosis of the abnormality researched on the embryo (Article L. 2131-4 of the French Public Health Code, last paragraph); if embryos are not "likely to be transferred or preserved" due to the expiration of conditions allowing the parents to procreate[18]; or if the parental project voluntarily ends (Article L. 2141-4 of the French Public Health Code). The couple's consent for research must be expressed explicitly in writing (Article L. 2151-5 II of the French Public Health Code) and, in the last case, "confirmed after a reflection period of three months" (Article L. 2151-5 II of the French Public Health Code). The disagreement between them or the absence of response and explicit consent lead to the destruction of the embryos (Article L. 2141-4 of the French Public Health Code).

Formal Requirements Regarding the Couple The delivered information must focus on the alternatives (donation to another couple or destruction) and on the consequences of the donation for research (destruction of the embryo, Article R. 2151-4 of the French Public Health Code). The information must also present "the nature of the planned studies in order to ensure [the couple] a free and informed consent" (Article L. 2151-5 II of the French Public Health Code). The article provides that "the derivative cells from embryos can be included in a preparation of cellular therapy for exclusively therapeutic purposes" (Article R. 2151-4 of the French Public Health Code). Nevertheless, the fact that cells might be appropriated by industry and become medication is eluded.[19]

Some Remarks Access to elements is simplified when it is a question of bodies without legal personality or when elements have already been removed. The notification is required but consent is not necessarily mandatory. Elements can be collected and subjected to a change of purpose without necessarily being part of a rigorous research protocol. Above all, for every single removal, whatever the modalities of consent (wirtten consent or mere absence of opposition), and the notification never specifies the nature of the organizations likely to collect them. Yet, samples from deceased persons for scientific purposes can be requested by commercial companies for their research programs. In the same manner, the consent for embryo donation must be free and informed, but no one is informed that the

---

[17]Article L. 2151-5 II of the French Public Health Code, confirmed by the prohibitions of Article L. 2151-2 of the same code.

[18]Article L2141-10 of the French Public Health Code, which refers to the conditions of the title, therefore notably Article L. 2141-3, or death, Article L. 2141-4.

[19]Prohibition by the European Court of Justice (ECJ) as well as the European Patent Office (EPO) regarding their patentability (ECJ, gr. ch., October 18th, 2011, Brüstle c/ Greenpeace, n°C-34/10 and EPO, February 4th, 2014, Technion research and Development Foundation LTD, T. 2221/10), and not the fact to seize them in the context of the medication regime (as testified by the Regulation EC no. 1394/2007 of the European Parliament and of the Council of November 13th, 2007 on advanced therapy medicinal products, cons. 7).

donation might be governed by the regime of medications for cellular and innovative therapies. Moreover, no one consents to the potential export of the collected elements (Chioccarello 2014, p. 257). The notification on the future of preserved elements in research involving the human person encounters the same flaws.

16.2.2 The Subsidiary Resource: Samples on Persons

Plan Article L. 1243-3 of the French Public Health Code derogates the regulation of the banks from the general provisions when "the activities regulated by the first paragraph [are] ensured in the context of research involving the human person." It therefore highlights that they are "regulated by specific provisions for these researches." It appears that the establishment of a collection when it affects the integrity of persons, is now always regulated by the Jardé Law. Several provisions authorize this interpretation. Yet, two regimes for sampling on persons exist under the Jardé Law: "classical" biomedical studies (1) and interventional studies presenting minimal risks (2).

16.2.2.1 In the Context of a "Classical" Biomedical Study

A Possibility to Sample All Elements When samples are collected during a biomedical research project, they are regulated by the overall proceedings applicable to biomedical research (site validations, mandatory approval of the Committees for the Protection of Persons since 2004, implicit "authorization" from the administrative authority, etc.).

In 1994, the legislative power authorized *tissues and cells* removals on a living person for scientific uses (Article L. 672-4 of the French Public Health Code), already applicable since 1993 for blood and its components as provided in Article L. 666-8 (last paragraph). However, it requires the direct therapeutic purpose for organ removals on the living (Article L. 671-3 of the French Public Health Code). However, a Decree of April 1st, 1997, created Article R. 671-8 whereby "for [organ removals], samples collected during biomedical research as defined by Article L. 209-1 are considered samples for therapeutic purposes."[20] All elements from the human body, including organs, might therefore be removed in the context of biomedical research (Chioccarello 2014, pp. 267–268). The only distinction concerns the authorization regime, which is stricter when an organ is sampled on a living person for scientific uses, regulated both by the authorization regime for biomedical research (Committee for the Protection of Persons and the French Agency for the Safety of Health Products) and by the regime applicable to samples for therapeutic

[20]Decree No. 97-306 of April 1st, 1997.

purposes (institutions allowed by the General Director of the French Biomedicine Agency, Article R. 1233-2 of the French Public Health Code). The same provision was used in Articles L. 1235-4 (organs) and L. 1245-6 of the French Public Health (tissues and cells).

All samples can therefore be collected for the profit of university institutes or pharmaceutical laboratories. Sampled elements are meant to be destroyed after the research, but their preservation is allowed if declared (in instances of a change of research purpose). Consequently, biomedical research can supply public hospital banks as well as those of pharmaceutical companies.

A counterpart to these provisions, the 2004 Law authorized sampling for establishing collections in a new simplified context, by allowing removals outside of any biomedical research protocol. This activity was reconsidered in 2012 and became classified as studies presenting minimal risks and constraints.

16.2.2.2 In the Context of a Simplified Regime

Non-biomedical Studies, in 2004: A New Provision for Blood The 2004 Law introduced Article L. 1221-8-1 which enables blood samples "for a purpose of establishing a collection of human biological samples" in non-biomedical studies. The article refers to the general regime of blood donation, both in terms of consent and volunteerism (Article L. 1221-3 of the French Public Health Code) and in terms of sanitary safety (Article L. 1221-4 of the French Public Health Code). The referenced blood can also concern blood subjected to modification of its characteristics (Article L. 1221-6 of the French Health Public Code) unlikely to include "negligible risks," as the article requires. Risks can, however, easily be considered as negligible: neglecting them is enough. The purpose of these provisions was to create, in 2004, a new category of samples for scientific purposes whose procedure is less constraining: research on samples (Lamaire 2007, p. 33). Since a sample does not constitute human subject research, and the research will be conducted on previously collected samples, the regime has escaped the Huriet Law provisions since 2004. Nevertheless, the Committee for the Protection of Persons must give an opinion on these collections (Article L. 1123-7 of the French Public Health Code).

The Order of January 12th, 2009, confirmed in 2016, specifies that samples for non-therapeutic purposes can concern "all kinds of labile blood products and samples from an allogeneic blood donation, collected by the Blood Transfusion Centre."[21] These are for studies on separated elements (in laboratories), for manufacturing various products not subject to non-profit or education. From a practical point of view, nobody can witness this turning point: the Order provides that the reception procedure for the donor is identical to the procedure during a blood

[21]Order of January 12th, 2009 instituting selection criteria for blood donor, Official Journal of 01/18/2009, p. 1067 replaced (but the provision remains by Order of April 5th, 2016, Official Journal of 04/10/2016.

donation. The question of scientific use, is included among the other classical questions in the pre-donation questionnaire.

This is therefore a new ground for infringement of physical integrity, outside of research involving human subjects, approved in 2004. Being sampled was possible, outside of any research program, for "commercial companies" (mentioned by the 2007 Order) to establish collections.[22] According to Mr. Dupont, in creating this new category of studies, the legislative power did not consider limiting it only to blood: the establishment of sample collections would be possible "for other components or products originating from the human body other than blood" if risks are negligible (Dupont 2008, note 32, p. 54). This approach has effectively been followed.

The Return of These Samples to the "Minimal Risks" Biomedical Studies In 2012, the legislative power removed the provision related to blood, and reintegrated these samples in the context of the general regime for research, no longer *on* the human person, but *involving* the human person (Article L. 1221-8-1 of the French Public Health Code). All research involving the human person are included (even if they do not infringe physical integrity), however, they are not all submitted to the same regime. Samples, collected not *on the occasion of*, but *for research* are therefore qualified as "interventional studies presenting minimal risks and constraints." In this context, research is not necessarily conducted by a doctor (Article L. 1121-3 of the French Public Health Code), does not have to be authorized by the French Agency for the Safety of Health Products, but only declared, the sole approval of the Committee for the Protection of Persons being required (Article L. 1121-4 of the French Public Health Code), nor does it have to be carried out in an authorized site, not even in a treatment facility (Article L. 1121-13 of the French Health Public Code). Under the new simplified regime, two types of research allow for the gathering of a collection.

Firstly, collections established by samples collected for this purpose in the context of interventional studies presenting minimal risks: The Order of May 3rd, 2017[23] abrogating and substituting the Order of December 2nd, 2016[24] removed several samples intended to establish collections when they were included in research on routine care (blood samples from arterial puncture, cerebrospinal fluid collection, etc.). These collections can be established based on blood tests, superficial skin biopsies, tissues samples or biopsies extended or additional during treatment (the legal text does not appear coherent here), urine collection, etc. The Order regulates the amount of blood that can be drawn in this context, implying that, above these quantities, samples should be included in the biomedical research of Category 1. Moreover, more invasive sampling initially planned in the 2016 Order do not seem impossible.

[22]Order of August 16th, 2007, Official Journal of 08/18/2007, p. 13825, annex 1.

[23]Order of May 3th, 2017 instituting the list of the studies mentioned in 2° of Article L. 1121-1.

[24]Order of December 2nd, 2016 instituting the list of the studies mentioned in 2° of Article L. 1121-1, Official Journal of 12/06/2016.

Indeed, they would be considered as samples collected during studies on routine care (the so-called "additional tube"), which allowed, since before the 2012 Law, to establish collections.[25] This is the other aspect of preservation "in the context of a study" that the Jardé Law aims at, creating the possibility of establishing collections *on the occasion* of a biomedical research, but without any specific research protocol to establish the collection, and involving the competence of the Committees for the Protection of Persons.[26] The establishment of collections following a classic biomedical study (e.g. on medication, which does not justify a supplementary sample for research) or research during routine care which may result in supplementary infringements (but not declined ones) *considered minor*, seem to be targeted here. In both cases of sampling *on the occasion* of a research, previously collected samples are preserved and others are collected additionally but "at minimal risk," during routine care research. The Clinical Research Associate, laboratory or hospital, can then preserve elements once the study has been achieved.

It appears, in this context of new simplified regimes, that the protection of persons is not ensured. Even for a blood test, the risks are not minor: 5507 serious adverse reactions were recorded in the context of blood sampling in 2016.[27] They are therefore simply declared as such *because they are minimized*. Then, the administrative authority, the French Agency for the Safety of Health Products, has not to authorize the research for routine care[28] since 2004, and although the establishment of collections were declared to the administrative authority, this has no longer been the case since 2016.[29] Yet, the Inspectorate General of Social Affairs estimated only

[25]Indeed, the Order of May 3rd, 2017 states that "acts which, in the context of research, form part of a care routine" can be realized in this context. Yet, under the Order of January 13th, 2010 instituting the standard rules and regulations which must be adopted by the Committees of the Protection of Persons (Official Journal of 02/05/2010) "the committee is also consulted for research project *on care routine*, project for which the opinion, if negative, prohibits the implementation of research, in case of preservation and preparation for scientific purposes of tissues and cells from the human body, including blood and its elements, for the needs of its own research programs, this situation including the establishment of a collection of biological samples." The question is to determine if these are samples collected during the treatment (comparison between a pathogenic and a healthy tissue to confirm the diagnosis, or if these are additional samples, unjustified by the diagnosis or the treatment).

[26]The new Article L. 1123-7 of the Public Health Code conferring to the Committees for the Protections of persons the mission to give an opinion on the "scientific and ethical relevance of the projects for the establishments of biological samples collections during research involving the human person". It must be deduced that this project of establishment is not a research but that the person is sampled during this one to establish a collection.

[27]ANSM, *Rapport d'activité 2016*, p. 41.

[28]Law No. 2004-806 of August 9th, 2004, Art. 88.

[29]Art. L1123-12 from order no. 2016-800 of June 16th, 2016, Art. 3. Declarations to the Ministry of Higher Education, Research and Innovation and Regional Healthcare Agency is not applicable under Article L1243-3 of the French Health Public Code as it is, therefore, rules related to biomedical research, that are relevant to apply (as the article reminds). The expressed authorization provided for researches on organs and tissues, Article R. 1125-7, only applied for international studies.

1% of the protocols audited by the Committees for the Protection of Persons between 2001 and 2004 were refused (D'Autume and Roussille 2005, p. 24)—this was before the significant extension of the CPPs missions in 2004, which was followed by another extension of their prerogatives in 2012, with short deadlines and a glaring lack of resources.[30] At the time, the Inspectorate offered a several hypotheses to explain it and called on them to reflect on the fact that they "do not exercise their vigilance with sufficient severity," (D'Autume and Roussille 2005, p. 37) observing an important number of interruptions imposed by the administrative authority because of major risks not identified by the Committees.[31] Committees do not guarantee the protection of persons, while the administrative authority can do so. However, in this new regulation of collections, it is easier to sample than before and the safety of persons remains in the hands of the Committees.

16.3 The Rights on Elements Once the Bank Has Been Established

Once elements have been detached from their body, the person has no rights on them (Sect. 16.3.1), unless such absence is conditional to a variety of obligations of the owners who intend to use the elements exclusively for public health (Sect. 16.3.2).

16.3.1 The Almost Non-existent Rights for Natural Persons

Absence of Right to Withdraw on Residues Articles regulating the change of purpose for every element exempt from their regime the Article L. 1211-2 paragraph 1 of the French Health Public Code, which requires the consent and its revocability at any time (for organs, tissues, and cells, Article L. 1245-2 of the French Public Health Code; for embryonic and fetal tissues, Article L. 1241-5 of the French Public Health Code). Therefore, they confirm that consent or absence of opposition can only be collected at the occasion and at the same time as the initial sample, which refers to the provisions initially mentioned.[32] The only derogation implies germinal cells and tissues, which can be used for other purposes, except "in

[30]For this analysis in 2009, RIAL-SEBBAG E., *Implications juridiques. . ., op. cit.*, pp. 156–157 (Committees are "ethical box record"). In 2012, CPPs still benefit of complementary competences, in personal data matters. For a critical analysis of this extension, see MATTIUSSI J., "Entre simplification des procédures et protection des individus: le nouvel équilibre de la loi Jardé" (to be published).

[31]"For the General Direction of Health: 33 in 2001, 24 in 2002, 3 in 2003;—For the French Health Products Safety Agency: 28 adjournments of test medications in 2002 and 9 in 2003", IGAS, ibid., p. 37.

[32]Point I, A, 1. *above-mentioned.*

case of death of the subject" (Article L. 1211-2 of the French Public Health Code). However, it is less about the person's right to examine than an obligation for service.

Absence of Right to Restitution on Other Elements French tribunals refuse to confer upon any person a right on elements and products of the body after their detachment.

An absent right on the ground of property. Judges refused to return gametes on the basis of Article 544 of the French Civil Code and Article 1 Protocol 1 of the European Convention for the Protection of Human Rights, because "there is no patrimonial right over the gametes claimed by Mrs. D."[33] In another case, a widow claimed the—possibly temporary—restitution of the samples collected from the body of her husband, who had suffered from leukemia, for paternity testing. The Public Assistance-Paris Hospitals refused to return them, which was confirmed by the administrative tribunal. Subsequently, the Court of Appeal ruled that:

> Pursuant to the joined provisions, articles L. 1243-3 and L. 1245-2 of the French Public Health Code, tissues, cells and products from the human body sampled during medical examinations and surgical interventions can only, subject to the provisions of article 16-11 of the French Civil Code, be returned, whether or not in return for payment, to organizations authorized by article L. 1243-3 of the French Public Health Code.

Since no physical person is mentioned in the subjects prone to receive elements, these are never returnable. The exception concerns Article 16-11 of the French Civil Code (returned to the expert appointed by judges, which was not the case here).[34]

An absent right on the ground of access to the medical file of the patient. The decision was confirmed on the ground of access to the medical file. Under Article L. 1111-7 of the French Health Public Code, each person, and the eligible dependents under certain restrictions, "have access to all information regarding the patient's state of health," but judges refused to see in "the collected samples of organic matter [...] one of the formalized information that rightly belongs to the claimant ."[35]

The only exception concerns *legal autopsies*. After several rulings refusing the restitution of sealed elements,[36] the legislative power allowed it solely if it respects the public safety requirements and if "these samples are the only elements enabling the identification of the deceased." It is therefore less about samples than remains, as it was the only way to allow the family to organize obsequies.[37] However, these are

---

[33]Toulouse Administrative Court, October 13th, 2016, No. 1405903.

[34]Paris Administrative Court of Appeal, February 13th, 2008, No. 06PA02800.

[35]Paris Administrative Court of Appeal, ibid. and Paris Administrative Court of Appeal, February 13th, 2008, No. 07PA01917.

[36]Cassation, Criminal Chamber, April 3rd, 2002, No. 01-81.592; Toulouse Court of Appeal, April 28th, 2009, No. 2009-377243; Cassation, Criminal Chamber, February 3rd, 2010, No. 09-83468, Bull. crim., No. 18, p. 84.

[37]Article 230-30 of Code of Criminal Procedure created by Law No. 2011-525 of May 17th, 2011, Art. 147.

not available to doctors: under the current law, samples are destroyed once the proceeding is achieved.[38]

The More General Refusal for Recognition of Rights If the consent is, under certain circumstances, explicitly required, it does not justify the existence of rights, as already specified by the regime on fetal tissues. This was confirmed by the French Constitutional Council which considered that "the legislator's choice to condition the sampling of these cells on the obtention of the prior written consent of the woman did not, in its aim or its effect, confer rights to her on these cells."[39]

An Exceptional and Limited Withdrawal Right Therefore, the person @@ can neither control the use nor the return of sampled elements. However, the person is entitled to a withdrawal right in two situations. Firstly, in the context of biomedical research, the donor can withdraw at any time (Article L. 1122-1 of the French Health Public Code) and so, elements already collected seem unusable for publications (Chioccarello 2014, p. 209) even if no legal text officially requires their destruction (Bellivier 2009, p. 139). Secondly, in the context of an embryo donation for research, the expressed consent was initially revocable "at any time and without any justification," but the provision was criticized, especially "when the embryo generated a derivation of embryonic stem cell line, from which cells have been removed and are impossible to distinguish within the research material" (Conseil d'Etat 2009, p. 23). Since 2011, consent has been "revocable without any justification as long as the researches have not been launched" (Article L. 2151-5 III of the French Public Health Code).[40] The banks' possession of the embryo is therefore constrained because the couple is aware of this possession and can reconsider this donation. Nevertheless, no research program is likely *to stop* because of the revocability of an initial consent.

It must be noted that, if not the person sampled, that is those collecting and using these elements that have rights on these elements as soon as they respect the process relative to their use and the (minor) constraints with regard to the "supply persons."

16.3.2 The Rights of the Bank on the Elements

A Property Right A property right on the element arises once this element is detached, and since the individual from whom it was detached has never been its owner, the property belongs to a third party. The question appears less a matter of the existence of an owner than of his/her identity, the use, the transfer to a third party,

[38]For critics and reform of these provisions, see Rougé-Maillard et al. (2016), pp. 4–13.

[39]Constitutional Council, Decision No. 2012-249 QPC of May 16th, 2017, §7.

[40]Law no. 2011-814 of July 7th, 2011, Art. 41. Except in case of study (focusing on the improvements of medico-psychological treat AMP technics), R. 2141-21 from Decree No. 2012-467 of April 11th, 2012, in which case it remains revocable at anytime.

and the element's alienation being possible. Furthermore, recalling the Administrative Court of Appeal in the 2008 above-mentioned decision,[41] banks, under Articles L. 1243-3 and L. 1243-4 of the French Health Public Code, can only transfer elements to legal entities. Consequently, the person subject to sampling is never the owner, *but the researcher (physical person) establishing the collection can neither claim the ownership nor have physically returned "his/her" collection to him/her.*[42] The legislative power distinguished two regimes for banks (Articles L. 1243-3 and L. 1243-4 of the French Public Health Code), depending on the main destination of collections (use for their own studies or for cession).

Two Kinds of Banks *Banks established for "the needs of their own research programs" (Article L. 1243-3 of the French Health Public Code).* The organization preserving and preparing tissues, cells, organs, and blood and its components or derivatives is governed by a declarative regime to the French Minister of Higher Education, Research and Innovation and its "activities include the establishment and the use of collections of human biological samples." Each preservation site must be declared (Article R. 1243-50 of the French Health Public Code) just as every collection must be. The declaration is registered at the French Ministry of Higher Education, Research and Innovation, as well as at the General Director of the Regional Hospitalization Agency (RHA) if it is a healthcare institution, and to submit an opinion on it at a Committee for the Protection of Persons (Article R. 1243-51 of the French Public Health Code). The declared activity can be launched, without opposition, 2 months after the file is considered to be complete (Article R. 1243-54 of the French Public Health Code) but the Minister can at any point, when appropriate, with the General Director of the Regional Health Agency, suspend and ban the activity if the conditions are not respected (Articles L. 1243-3 and R. 1243-58 of the French Public Health Code).

The activity may give rise to two kinds of contracts, referring to transfers of elements and products from the human body, which are not considered legal cessions. Article R. 1243-60 of the French Public Health Code regulates transfers when the research involves several organizations and that, in this context, some elements are supplied to another research structure (at the occasion, for instance, of a program involving a collaboration between several structures); this also applies when an organization hires an external service provider. In both cases, the organizations return to the structure which had declared the activity all elements included in the contract. This is why transfer differs from cession.

Banks whose main activity is cession. In contrast, banks must be authorized (and not declared) when they preserve and prepare organs, tissues, and cells "for their cession" (Article L. 1243-4 of the French Public Health Code). The absence of

---

[41]Paris Administrative Court of Appeal, February 13th, 2008, No. 06PA02800.

[42]At the most it could claim an intellectual property rights on its discoveries, which appears to cause conflicts of norms, as the discovery in biotechnologies can be embodied in elements. See also Chioccarello A., *Le matériel biologique humain*, *op. cit.* p. 273.

approval from the Minister of Higher Education, Research and Innovation within 3 months constitutes an implicit decision of rejection (Article R. 1243-67 of the French Health Public Code). This article essentially intends to apply Biological Resources Centers, whose mission is to ensure the availability of annotated and preserved elements for the research community.

If these two kinds of contracts and partnerships seem to enable the circulation of elements, it is more difficult to understand the absence of requirements of not-for-profit structures, of the centralization of information on their resources, or of their possibility to escape the knowledge of all concerned.

The Absence of a Non-lucrative Requirement for Banks and Contracts on Elements Structures preserving for their own research programs can belong to both the public and the private sectors. Although the legislative power, in 1994, required non-lucrativity as a condition of authorization for the banks, regardless of their purpose—therapeutic or scientific (Article L. 672-10 of the French Public Health Code in 1994)—it waived this condition in 1996 when there is a genetic purpose, and then more generally for any scientific preservation.[43] Moreover, there is no fee requirement for contracts on elements, whether it is a case of transfer or cession. The term "cession" have taken on here a new meaning with regard to elements and products from the human body. Contrary to the use of "cession" in the "cession fees" invented with blood and applied to exclude any profit for transfer of property between organizations, of elements and products of the human body for therapeutic purposes, the scientific use of the term in includes all types of property transfers without requiring them to be free nor the bearing of the costs resulting from the entire activity. There are two ways to generate the circulation of elements without profit: the establishment of "cession fees" by the administrative authority or the funding of the activity by a public contribution granted to the institution for its operating costs.[44] There is no organization of this kind for the use of elements for scientific purposes, despite recommendations insisting on the fact that "biological materials should not be, as such, a source of profit." (Bioy 2013, pp. 147–152)[45] In this case, the retrieve of elements is possible for lucrative purposes (for the own uses of industry, on the occasion of transfers or cessions contracts) before any transformation. The hospital, in the context of a public-private partnership, might have to remunerate for the availability (for lucrative purposes) of a collection for research while it conditioned its establishment (through sampling).

The Absence of Centralization and Knowledge About Resources Finally, there is no optimization of the resources through their centralization. Firstly, because,

[43]Law No. 96-452 of May 12th, 1996, Art. 1, then Articles L. 1243-3 and L. 1243-4.

[44]Referring to our thesis developing these aspects, Catto M.-X., *Le principe d'indisponibilité du corps humain, limite de l'usage économique du corps*, published in LGDJ, 2017, §323–341 (for blood) and §343-352 (for gametes and organs).

[45]Rec (2006)4 of the Committee of Ministers of Council of Europe to Member-States for research using biological material from human, adopted in March 15th, 2006, Art. 7 of the annex. What it is however, in an upgrading logical potentially opposed to the general interest,

under Article L. 1243-3 of the French Health Public Code, "activities regulated by the first paragraph practiced in the context of a research involving the human person are governed by specific provisions for these studies." Yet, for these, and for collections established in this context, no obligation of declaration or authorization is required. The only constraint consists in an evaluation by the Committees for the Protection of Persons, of the relevance of the collections' establishment (Article L. 1123-7 of the French Public Health Code), the other provisions refer to Articles L. 1243-3 and 4 (for instance Article L. 1131-4) and relate to provisions on research involving human subjects. Therefore, there is a circle allowing the absence of declaration, at least before any new research protocol on preserved elements (under French law, an exportation could have, meanwhile, been authorized under Articles R. 1235-7 et seq. of the French Public Health Code.)

But even though the industry would submit a declaration, nothing can constraint neither the private sector nor the public one to make the established collections accessible to the public. Yet, if we admit the fact that the donors do not have any right of examination on detached elements, for any future use, because it would "result in the creation of insurmountable obstacles in scientific advancement" (Bellivier 2001, p. 58; Le Roux et al. 2003, p. 810) and that it is wiser to let the researchers research (Chioccarello 2014. p. 209) to the extent that "nobody can accurately plan for which specific study samples and data collected in biobanks will serve in the future," (Soulier and Cambon-Thomsen 2016) it appears that it *is possible only if elements are intended for research and public health.* Or, failing that and *as a minimum*, by turning the private activity into a public one. In other words, the reduction of ethics to the consent of subjects (when required) omits questions yet related to personal protection such as the optimal use of elements, their proper preservation, the centralization of information to avoid duplications, access to existing collections rather than establishing new ones, etc. (Soulier and Cambon-Thomsen 2016, p. 40). It is thus difficult to understand that all elements are not preserved by public and specialized resource centers. In this context, these centers would preserve elements that they own for their programs or ensure the availability of resources for other partners (public or private) by transfers or cessions. They could also be simple depositor banks for other public structures owners. A national directory could register every collection and another could register all ongoing research (Bioy 2013, p. 147 et p. 151). Although a directory has been planned for the second option by Article L. 1121-15 of the French Public Health Code (since 2012),[46] the decree relative to this norm has not yet been issued. Consequently, nothing prevents companies, which constitute gigantic banks, from "keeping them fiercely locked today." (Bellivier and Noiville 2009, p. 93).

The catalog of the set of elements would therefore be available to the entire scientific community, the quality of samples being guaranteed and invisible, private studies being avoided.

[46]For the former provision, that was not public and allows to block the registration by the ANSM for researches in various field, see Decree No. 2006-477 of April 26th, 2006.

References

Bellivier, F. (2001). Matériaux et résultats de la recherche en génétique humaine: richesse sans propriétaires?. *Droit et patrimoine*, (91), 58.

Bellivier, F. (2009). Le droit de retrait en bioéthique sur les voies de l'anticipation. *Droits*, (48).

Bellivier, F., & Noiville, C. (2009). *Les biobanques*. Paris: PUF.

Bioy, X. (2013). Les biobanques entre rareté et accès partagé aux ressources biologiques humaines. In *La santé publique à l'épreuve de la rareté* (pp. 147–152). PUAM.

Catto, M. X. (2017). *Le principe d'indisponibilité du corps humain, limite de l'usage économique du corps*. Paris: LGDJ.

Chioccarello, A. (2014). *Le matériel biologique humain: étude juridique sur l'utilisation à des fins scientifiques des éléments et produits du corps humain.* Thesis, dactyl., Paris V.

Conseil d'Etat. (2009). *La révision des lois de bioéthique*, étude adoptée par l'Assemblée plénière, Paris (p. 23).

D'Autume, C., & Roussille, M. B. (2005). *La transformation des comités consultatifs de protection des personnes en matière de recherche biomédicale (CCPPRB) en comités de protection des personnes (CPP) applying the August 9th, 2004 Law*, Rapport IGAS.

Dufourcq, E., Lalande, F., Delahaye-Guillochau, V., & Ollivier, M. (2002). *Conservation d'éléments du corps humain en milieu hospitalier*, Rapport IGAS.

Dupont, M. (2008). *Recueillir, conserver et utiliser des échantillons biologiques humains à l'hôpital.* APHP, Doin, Lamarre.

Harichaux, M. (1988). Le corps objet. In *Bioéthique et Droit*. Paris: PUF.

Hauw, J. J., & Ardaillou, R. (2009). *Rapport relatif aux modifications à apporter à la législation sur les centres de ressources biologiques.* BAM.

IGAS. (1993a). Inspection Générale des Affaires Sociales. In *Enquête sur les banques de tissus d'origine humaine.*

IGAS. (1993b). Inspection Générale des Affaires Sociales. In *Rapport d'étape sur les banques de tissus d'origine humaine.*

Lamaire, F. (2007). Recherche biomédicale: évolution récentes. In *Essais cliniques. Quels risques?* Paris: PUF.

Le Roux, N., de Montgolfier, S., di Donato, J. H., Boccon-Gibod, L., Teillac, P., Hervé, C., et al. (2003). Les collections de ressources biologiques humaines pour la recherche: de la réglementation à la mise en place d'un guide de bonnes pratiques du collectionneur. *La Revue de Médecine Interne, 24*(12), 804.

Memmi, D. (2011). *La seconde vie des bébés morts*. Paris: EHESS.

Mesnil, M. (2017). L'embryon, l'alpha et l'oméga de la recherche. *Journal de la Santé et de l'Assurance Maladie*, (special No. 15).

Rial-Sebbag, E. (2009). *Implications juridiques des nouvelles formes de gouvernance en biotechnologie: l'exemple des biobanques utilisées en recherche.* thèse, Toulouse (p. 128).

Rougé-Maillard, C., Cariou, A., Jousset, N., & Bouteille-Brigand, M. (2016). La recherche scientifique sur les prélèvements biologiques réalisés dans le cadre des autopsies et objets de scellés judiciaires. *Médecine and Droit*, 4–13.

Soulier, A., & Cambon-Thomsen, A. (2016). Promesses de biobanques: se soucier de l'avenir dans l'éthique de la recherche biomédicale. *Revue française d'éthique appliquée, 2*(2), 29–47.

Chapter 17
The Patrimonial Specificities of Public Persons

Benoît Schmaltz and Pierre Égéa

A public person, who is "genetically carrier of general interest" (Yolka 2013, p. 13), is only the owner to fulfil their role, their missions and reach goals that are associated to it. Their relationship to things is immediately and consistently finalized, and, in these two areas, goods which they own are subject to the specific arrangements of the resulting public property (Yolka 1996). That's why a public person is not owner as private individuals are. Their assets are public and no other criterion than the public figure of the owner is required (Chamard 2004, n°895: "Public goods can then be defined as the property owned by legal persons of public law") as well as their right to property, which is public—not private as that of Article 544 of the civil code—the exercise of which is immediately assigned to the satisfaction of the legitimate interests of their role (Schmaltz 2016). Therefore, specific patrimonial features which, certainly, have such a level of generality that their scope seems limited (Gaudemet 2014; Sorbara 2016) or their independence uncertain (Chamard 2004), are no less essential.

It is through the prism of this public property right that tumor banks and biobanks will be approached. However, we shall exclude the aspects that are the object of specific communications. It is assumed that the movable assets of biobanks are part of the private domain.[1] It will also be assumed that samples, data, collections and biobanks may be the property of a public person, without going back on the issue of the nature of assets for tissues and elements of the human body, nor on the issue of their appropriation.[2] Finally, we shall voluntarily exclude the issue of common assets which would lead to a substantive general discussion on this heterodox

[1]See Mr. Maxime Boul's speech.

[2]See Ms. Marie-Xavière Catto's speech.

B. Schmaltz (✉) · P. Égéa
University Toulouse 1 Capitole, Maurice Hauriou Institute, Toulouse, France
e-mail: benoit.schmaltz@ut-capitole.fr; Pierre.Egea@ut-capitole.fr

X. Bioy (ed.), *Public Regulation of Tumor Banks*,
https://doi.org/10.1007/978-3-319-90563-1_17

theory.[3] Therefore, tumor banks and their collections will be considered as public assets, thereby objects of the ownership right of public persons.

A public owner would more or less benefit from the exemptions which characterize administrative law (Rivero 1947, p. 69). Notably, their property will be intangible and will not be the object of operations at a low price, which would be synonymous with liberalism. Such is the case for constituent assets of a tumor bank.

The latter can be considered as a standalone asset. Indeed, the technique of universality helps to achieve this result. Universality can be defined as "a new asset, composed of autonomous elements driven by a legal assignment community" (Denizot 2008, p. 152). Thus, this technique first allows one to universalize samples, data and everything that relates to them and to define the collection as an independent asset from these elements. It is the same logic as the one presiding over the identification of the collections in museums or books in libraries.

The universality technique then goes further still to represent by such an asset the entirety of a tumor bank or a biobank. It would therefore be the representative universality of an activity, such as a business or any other professional fund well known in private law. If the comparison may surprise or shock, it is quite an interesting approach: just imagine a tumor bank as an administrative fund, assigned to the general interest. Thus, it would constitute a standalone unit, and tumor banks and, more widely, biological resource centers, become constituent elements of a public service of the biobank, which largely remains to be built.

Thus, patrimonial specificities of public persons are twofold: those regarding public property (Sect. 17.1) and, in addition, those within the public service (Sect. 17.2).

17.1 Specificities Regarding Public Property

As far as public property is concerned, the privileges and the constraints of administrative law are, on the one hand, an increased protection of public assets because of their exemption from seizure (Sect. 17.1.1) and, on the other hand, a limited flexibility in their management owing to their non-assignment at a low price (Sect. 17.1.2).

17.1.1 An Increased Protection: The Exemption from Seizure

Article l. 2311-1 of the general code on public property states that: "the property of public persons referred to in article L. 1 are exempt from seizure.". As a legislative

[3]See Ms. Florence Bellivier and Christine Noiville's speech.

transcription of a general principle of law recognized by the Court of Cassation,[4] it is a fundamental rule of administrative law of property. It has been criticized because of its potential clash with competition law, notably that of the European Union (Roux 2015), but it has nonetheless been reaffirmed and has not, so far, been subject to direct exemption. The result is that a public person is immune, in general, against any civil enforcement. They will not therefore be deprived of the free disposition of property which they own. This means that the property of a public person is free from any seizure whilst they themselves cannot be subject to judicial collective procedures, recovery or liquidation.

This rule is of interest to biobanks and tumor banks. All the elements of a tumor bank are safe from the creditors of the public person that owns it. However, the exemption from seizure does not protect a tumor bank against a decision of their owner to dismantle it. Indeed, the exemption from seizure should not be confused with the inalienability which only relates to the public domain and which, moreover, is a more procedural than substantial rule as it is actually possible to get around it. Public property offers no response to the fear we have been able to read about here or there of a dispersion of the elements of a biobank (Bellivier et al. 2009, p. 122), apart from the action of creditors.

One should also mention the existence of an administrative procedure in which the prefect may, to pay the debts of a public defaulter, automatically decide the sale of certain assets, "if they are not essential to the functioning of public services they are responsible for" the Council of State[5] fortunately states. This could well be the case for a tumor bank and its elements, but this hypothesis calls for some attention.

The scope of protection that is the exemption from seizure is therefore relatively limited. It is a little different with the constraint imposed on public owners regarding the consideration of the operations performed on their property.

17.1.2 A Limited Flexibility: Non-assignment at a Low Price

A public person is already subject to several controls that affect them as owner. As a public manager by nature, a public person is first subject to the control of financial jurisdictions. As a contracting authority by nature as well, a public person is subject to the public procurement law, and that situation should not be ignored by tumor bank managers. Indeed, they are often qualified as contracting authorities within the European Union's law, and so they are likely to conclude public procurement

[4]1st Civ., 21st December 1987, *BRGM c/ Lloys Continental*, *RFDA*, 1998, p. 771, B. Pacteau's note, Charbonnier's conclusion; *CJEG*, 1988, p. 107, L. Richer's note; *JCP G*, 1989, II, no. 21183, B. Nicod's note, *RTD Civ.*, 1989, p. 145, R. Perrot's note; *GDDAD*, *Paris, Dalloz*, 2013, p. 663, Ph. Yolka's note.

[5]Council of State, 18th November 2005, *Société fermière de Campoloro et autre*, Req. No. 271898.

contracts for services or supply or, also, service concessions (public or not) when a third party is authorized to exploit a collection for their own benefit.

Finally, and this is what directly concerns us here, a public owner is subject to the general prohibition of consent to donations. As a traditional public law rule,[6] it has been upheld by the Constitutional Council for which constitutional protection of public property "opposes the possibility that assets or companies, which are part of public property, be transferred, at prices below their value, to persons pursuing private purposes".[7]

Applied to the field of property, this rule has been worded as an anti-assignment at low prices of the assets of public persons.[8] The jurisprudence is now relatively profuse[9] to determine if compensation, monetary or not (which is trickier) is sufficient to prevent grievance. It emerges that the property of a public person cannot be transferred to a person pursuing a private interest at a price below its market value without counterparties of general interest justifying this difference.

However, the general scope of this rule should be noted. It is not only the sale of a public asset at a price below its value which is, in principle, prohibited, but also any legal monetary operation that is performed on this property. Thus, the price of rental or use of a public good is also subject to the anti-assignment rule at a low price. It explains the jurisdictional control operated on the amount of state fees.[10] Under the requirements of Article l. 2125-3 of the general Code of public persons' property, the amount of the fee must be determined by considering the "benefits of all kinds" provided to the occupant. An insufficient amount would therefore justify the annulment of the decision which secured it.

This logic should have its counterpart in the private sector by challenging a price too low charged to the other party of a public person (*contra* Moysan 2001, p. 249). In fact, private domain management obeys rules of public law whose scope ignores the limits of legal and jurisdictional dualisms. Thus, the principle of equality applies, and it is opposed to reserving the free use of hunting rights to an individual,[11] and has also resulted in significant litigation regarding personnel accommodation (Duroy). In general, it seems legitimate to consider that any legal transaction on the property of a public person must have sufficient consideration, either monetary or of general interest.

---

[6]Council of state, 17th March 1893, *Chemins de fer de l'est*, D. 1894, p. 119, Romieu's conclusion; Council of State, 6th March 1914, *Synd. de la boucherie de la ville de châteauroux* : Rec. Council of State, p. 308.—25th Nov. 1927, *Sté des établissements Arbel* : Rec. Council of State, p. 1114.

[7]Constitutional Council, decision no. 86-207 DC of 25th-26th June 1986, recital 58, Rec. Const. coun., p. 71.

[8]Notably, following Mr. Philippe Yolka's PhD, aforementioned.

[9]From the decision of CS, sect., 3rd Nov. 1997, *Cne Fougerolles*, Req. No. 169473.

[10]CAA Paris, 17th Oct. 2013, no. 13PA00911, *Ville de Paris, Fédération française de tennis* : *JCP A* 2014, 110; *AJDA* 2014, p. 31, Mr. Sirinelli's chronicle.

[11]Council of State, 14th Sept. 1994, *Cne Escoutoux* : Rec. CS 1994, p. 413.

Therefore, biobank managers are concerned whenever consent gives access to the bank or the use of collections and samples. In principle, there should be a satisfactory return. Where access to biobank can be a source of profits, so the billing must be fixed at a price considering, at least, the cost for the public person and possibly the benefits expected by the user. Of course, we acknowledge that, regarding access to biobanks, biomedical research is obviously an activity of general interest to justify a tariff policy to encourage the broadest access. It is therefore essential to make a distinction, although the limit is sometimes difficult to identify.

Protected from seizure, but having to manage their assets for the general interest and with the concern of public finances, a public person operates their assets for public service activities, as tumor banks do. A public property law approach can contribute to the development of such a public service, more generally that of biobanks.

17.2 The Specificities Regarding the Public Service of Biobanks

As a first step, our approach to public property law can be deepened by applying the technique of universality to tumor banks, which allows one to design them, when they belong to a public person, as an administrative fund (Sect. 17.2.1). The exploitation of such a fund being a public service, it will be, subsequently, essential to define the stakes of a public service of biobanks (Sect. 17.2.2).

17.2.1 The Techniques of De Facto Universality and "Administrative Fund" in the Service of Tumor Banks

According to Xavier Bioy, "legal texts do not use the term "biobanks", they merely legislate by "strata" on the object kept by the bank, then on the opening of the bank or its action on the French territory when it is located abroad, and on the circulation of the data collected thereafter. These three elements make up a "biobank" without being assembled in a legal concept announced as such. The interest would be, for analysis as for public policy concerns, to move beyond the collection to reach biobanks, i.e. adding the institutional dimension: infrastructures, the organization (of the premises, equipment, dedicated people, safety devices), in short, a status. It would be beneficial to separate collections and biobanks, the first being the subject of the second" (Bioy 2010, p. 885). The technique of universality, in the form of a fund, can meet this expectation of a fund which would be administrative in public tumor banks.

Indeed, based on the model of professional funds, the most famous being the business model, it is possible to design a biobank as a fund. This is where using the technique of de facto universality consists in creating "a new asset, composed of autonomous elements driven by a legal assignment community" (Denizot 2008, p. 152). Thus, a set of books or works of art can become the collection of a library or a museum. A database or a collection of samples are also likely to be reunited in de facto universalities. The immediate interest, in such cases, is to carry out legal operations on the universality and not on each of these numerous elements. The first use of the technique of de facto universality in the case of tumor banks thus concerns these relatively homogeneous elements that can be combined into a single asset.

The technique of the fund is of a higher degree of complexity. Indeed, funds are a universality combining various things brought together by the fact that they are all assigned to the same activity (Chatain-Autajon 2006, p. 430). We can consider that it is about representing an activity by all the means that are assigned to it. Thus, Roman antiquity has known its first universalities of economic fact in the form of shops and agricultural funds: *taverna instructa* and *fundus instructus* (Chatain-Autajon 2006, pp. 33–43). It is business that, in modern French law, has opened the series of de facto universalities of this nature[12]: agricultural funds,[13] artisan funds,[14] liberal funds.[15] It is an approach that could be criticized in doctrine, and, besides, it has proposed a single notion of professional funds. The fact is that the technique of the fund, as any de facto universality, has a strength and a weakness which is its plasticity that allows an application to any activity with a repeatedly adaptable regime (Denizot 2008, p. 29).

This plasticity of the funds technique has been used to suggest its transposition in the field of public law in the form of the "administrative fund", defined as the universality of property assigned to an activity undertaken by a public person (Schmaltz 2016, pp. 331 et sq.); in short, any tumor bank can be formalized in the form of a fund. When a biobank is implemented by a public person, it is an administrative fund. In all cases, this de facto universality may in fact be analyzed from three angles: "in its entirety, from the point of view of its elements and from that of the links between its entirety and its elements" (Schmaltz 2016, p. 24).

As a whole, i.e. the fund itself, the main interest is to constitute the activity as being itself the subject of legal operations. Regarding professional funds, it is mainly the transfer, lease management implementation and security. An administrative fund will, for its part, be the object, for example, of a transfer of competence or a concession. But to respond to Xavier Bioy's solicitation to consider the 'status' of

[12]From practice, apprehended initially by tax law, business was explicitly established for the first time with the law of March 17th, 1909. It has been followed by others.

[13]Law no. 2006-11 of 5th January 2006.

[14]Law n° 96-603 of 5th July 1996, whose Article 22 authorizes security according to business rules.

[15]Recognized with decision Cass. 1st civ., 7th Nov. 2000, *Bull. civ.*, I, no. 273, p. 183; *JCP G*, II, 10452 note by F. Vialla; *Defrenois*, 2001, juris., p. 431, note by R. Libchaber.

a biobank, it is possible to associate to each fund regulations governing the modalities of its operation. It is at this level that we have general rules concerning the build-up of collections for use by third parties, either through a logic of broad access or without neglecting the development that these third-party uses can allow.

At the level of the elements, i.e. assets which are incorporated into the fund because of their assignment to the activity, the logic of superposition of this technique should be noted: the legal regime applicable to an element does not disappear with the allocation to a fund. Simply, the allocation to a fund, and especially an administrative fund, can have the effect of imposing on some or all of the assets affected the application of special rules. By reference to public property, we think immediately of inalienability, but one could also think of simple rules of control over the provision acts performed on the elements of such a fund.

As for the link between the whole and the elements, it is probably on that issue that allocation logic should apply in full. The administrative fund allows one to consider the theory of returnable assets as a general mechanism intended to apply to all administrative activities even outside the mechanism of concession (Schmaltz 2016, No. 484 et seq). By combining the theory of returnable assets with control of the acts of provisions, the following insoluble problem posed by the exemption to seizure would be solved: a dispersion of the fund by setting strict rules concerning the possibility of getting out of an administrative fund of this nature, such as samples or anything else that the importance of which would justify.

However, all biobanks are not administrative funds. Indeed, the latter are necessarily the property of a public person as they serve their role. A tumor bank or a private biobank is a professional fund of private law. Therefore, the question of differentiation of legal regimes arises between biobanks according to whether they are, or not, means of public action or the result of private initiative. The technique of de facto universality allows one to separate the regime applicable to any biobank and the regime only applicable to administrative funds.

This has enabled the envisioning of a network of tumor banks, other biobanks and all types of biological resource centers whilst confronting them with the concept of public services—that is a biobank public service—challenges of which it is now necessary to identify.

17.2.2 The Challenges of the Biobank Public Service

The concept of funds is part of the institutional reflection on biobanks by providing a legal tool for their conceptualization, including a public biobank by administrative fund. It is now a matter of confronting the concept of public service, first with biobanks taken in isolation, then with all biobanks, public and private.

If biobanks, taken in isolation, are confronted with the concept of public service, the distinction between public and private biobanks should be redefined. The main interest of the concept of administrative funds was to embrace all assets of the administration, whether they belong to the public domain, to the private domain or to a private person. However, to maintain the link to ownership of public persons and, with it, the organic unit of public action, the criterion by which a fund is classified as administrative or not has been identified as the property of the public person. It was this appropriation relationship which seemed to be the most able to account for the public law relationship between a public person and the activities within its purview, assuming an activity, for a public person, therefore means being the owner of an administrative fund, irrespective of a possible delegated management of the activity and of a private property for some of the affected assets. This criterion of public appropriation of the fund is nothing more than a patrimonial variation of the fundamental criterion of the qualification of a public service activity: the organic connection to the State and to public persons. It refers to the distinction between public services and activities benefitting the greater good (Plessix 2016, no. 658–671), i.e. the fact that the same activity will be a public service or a private activity depending on whether it has been assumed (taken over) by a public person whose capacity is recognized (Truchet 1982, p. 427).

Applied to biobanks, this means that the public service is a qualification that only overlaps with some of them. It is therefore necessary to distinguish between the applicable regime, because it is a biobank, and the regime applicable because it is a public service. For the most part, a biobank being a biobank, it should be governed as such. This is indeed why the legislature in Articles L. 1243-3 and -4 of the Code of Public health, which targets "organizations" and "health institutions", without specifying whether they are public or private submits them to the procedures of declaration or authorization. Just as private and state nursery schools are subject to identical obligations in terms of security, but only the latter must respect the principle of secularism,[16] one can imagine that some rules will apply only to biobanks attached to a public person, and thus qualified as public service.

Therefore, designing the rules applicable to biobanks does not necessarily mean designing a public service regime for biobanks. However, such a regime may be useful in relation to an activity that combines both the protection of personal data and important issues regarding public health and biomedical research. The normative scope of the public service qualification should nevertheless be confined to the enactment of the only rules which are necessary because of the principles generally governing public institutions or are imposed because of public service missions specifically implemented by public biobanks. In this case, a reciprocal movement of convergence between public and private biobanks can be envisaged, for example through the mechanism of association to the public service, well known to the health sector.[17] However, this extension of the public service does not seem to be on the agenda.

---

[16]See the "Baby Loup" case and the judgment of 19th March 2013 by which the Court of Cassation states that "the principle of secularism established by article 1 of the Constitution is not applicable to employees of private employers who do not manage a public service".

[17]C.C., 6th July 2009, no. 2009-584 DC, *Loi portant réforme de l'hôpital et relative aux patients, à la santé et aux territoires*, *RFDA* 2009. 1270 A. Roblot-Troizier's review; see Articles L. 6112-1

Neither public biobanks taken in isolation, nor even the consequences of an association to the public service depletes the public service of the biobank which is wider than their mere addition. It includes, first, the state authorities which actually oversee all the biobanks (Ministry of Health, Ministry of Research, ARS, etc.). Only then should only the biobanks that are under the control of a public person, health institution and/or research be included. It must be considered that these public biobanks are part of the public service of the biobank, primarily the responsibility of the aforementioned administrative authorities.

This public service is therefore not constituted by biobanks. Biobanks are the object of the public service. That is why a public biobank is first subject, in principle, to the common law of biobanks, and it is only then that the public service of the biobank should be added to their regime, in the service of two functions.

First, a regulatory function: whether it be public or private biobanks, the public service aims to establish the applicable legal regime and to ensure its regulation and control. This function shall be linked to the receipt of declarations and the issuance of authorizations to manage a biobank. During our reflection of public heritage law, one could add a right of pre-emption to any human biological material, collection or biobank that would be the subject of a cession.

Then a function of coordinating biobanks, in particular to meet the imperative of their networking at a national, European and global level. In this regard, one could imagine a public power prerogative to impose access to a biobank for a research project. This would only be a mechanism comparable to expropriation or servitudes established for public utility.

In conclusion, an institutional reflection on biobanks has at least three identified levels: the formalization of all biobanks in the form of universality and their submission to a common system; the formalization of public biobanks in the form of an administrative fund involving the application, by superimposition or by derogation, of special rules; and the formalization of a public service the purpose of which being all biobanks, implemented by the administrative authorities and public biobanks to satisfy, at best, the interests of health and research.

References

Bellivier, F., Morin, A.-L., & Noiville, C. (2009). Les biobanques s'invitent au banquet de la bioéthique : mode d'emploi. *Revue politique et parlementaire, no. 1 to 50.*

Bioy, X. (2010). Vers une politique publique des « biobanques ». *RDSS,* 885.

Chamard, C. (2004). *La distinction des biens publics et des biens privés, contribution à la définition de la notion de biens publics.* Dalloz, NBT.

Chatain-Autajon, L. (2006). *La notion de fonds en droit privé.* Litec.

Denizot, A. (2008). *L'universalité de fait.* Fondation Varennes.

et seq. of the Public Health Code defining the public health service and the empowerment of private institutions to operate it.

Duroy, S. JCl. Propriétés publiques, fasc. 79-60 Occupations privatives du domaine privé.
Gaudemet, Y. (2014). *Traité de droit administratif*, t. 2, *Droit administratif des biens* (15e ed.). LGDJ.
Moysan, H. (2001). *Le droit de propriété des personnes publiques*. LGDJ.
Plessix, B. (2016). *Droit administratif général*. LexisNexis.
Rivero, J. (1947). Droit public et droit privé : conquête ou *statu quo* ? *D.*, 1947, chron. XVIII.
Roux, C. (2015). *Propriété publique et droit de l'Union européenne*. LGDJ.
Schmaltz, B. (2016). *Les personnes publiques propriétaires*. Dalloz, NBT.
Sorbara, J.-G. (2016). *Manuel de droit administratif des biens*. PUF.
Truchet, D. (1982). Nouvelles récentes d'un illustre vieillard : Label de service public et statut de service public. *AJDA*, 427.
Yolka, P. (1996). *La propriété publique. Éléments pour une théorie*. LGDJ.
Yolka, P. (2013). *Les grandes décisions du droit administratif des biens*. Dalloz.

Chapter 18
Biological Sample Collection in the Era of Genomic Medicine: A New Example of a Public Commons?

Christine Noiville and Florence Bellivier

More than ever the medical sciences and the arts of prevention and care must combine individual aspirations and collective interests. While modern medicine has always used bodies, in more or less licit ways (Chamayou 2008) to advance medical knowledge, the last quarter of the twentieth century and the beginning of the twenty-first have been marked by four developments that have converged to make everyone, whether sick or healthy, an object of existing or envisioned medical research. First, live human samples (tumors, blood, gametes, urine, cells, etc.) were assembled into collections of biological samples—biobanks, which are infrastructures composed simultaneously of all sorts of samples and of data intended to illuminate the samples in as much detail as possible and serve researchers as tools for their projects. Next, genetics, although its specialists themselves stress that it is far from explaining all our behaviors and our diseases, continues to develop and remains fundamental for understanding the variability between individuals and, hence, diseases. Moreover, the conception of research on humans has changed over time, moving from a paradigm that clearly distinguishes between basic research and medical care to what is called translational research, intended to ensure that scientific and technical innovations and discoveries can be rapidly transferred from basic research towards applied research and patient care.[1] Finally, advances in computer

The authors thank F. Le Thimonier, A. de Guerra, A. Cambon-Thomsen, H. Esperou, C. Bourgain, F. Nowak, P. Boucher and Tom Billins for their comments and their counsel, and Jo Ann Cahn for the translation.

[1]On this theme, see the special report in issue 5 of *Cahiers Droit, sciences et technologies*, 2015, entitled "Les frontières entre recherche et soin", available at: https://cdst.revues.org/386.

C. Noiville (✉)
French National Center for Scientific Research, UMR 8103, Paris, France

F. Bellivier
University of Paris Nanterre, Paris, France
e-mail: florence.bellivier@u-paris10.fr

X. Bioy (ed.), *Public Regulation of Tumor Banks*,
https://doi.org/10.1007/978-3-319-90563-1_18

science have led to methods of handling what is called *big data*. These methods are applied, for example, in genetics to deal with the information acquired in the very rapid sequencing of whole genomes.[2]

The fusion of these aspects of scientific evolution—or revolution—are leading to the increasing individualization, or precision, of medicine, in the sense that substances are developed to treat a given disease in a given individual. To do this, however, research must be increasingly more collective, for the consolidation and amplification of knowledge depend on the mass of data collected and their subsequent comparison. The quality depends, among other things, on the quantity. At the same time, the interrelationships between individuals, physicians, researchers, and society are growing and lend support to the idea that, essentially, health is everyone's responsibility, especially in a constrained, financial environment. This concept, in turn, raises serious questions of social justice. That is, "[i]f precision medicine means that we should all be treated differently, how can we satisfy the principles of justice and fairness that require that people are treated the same?" (Mills 2016).

Consequently, the model of the *commons* appears theoretically relevant for thinking, not so much about health in general—health as an international public good has been the subject of numerous studies (Boldin 2015)—but about the governance of the health tool that biobanks constitute, especially as we enter the era of genomic medicine (Sect. 18.1). Nonetheless, the governance of biobanks, as transformed by genomic medicine, will have to meet some conditions in actual practice to constitute a true commons (Sect. 18.2).

18.1 The Commons, a Theoretically Relevant Model

As a prerequisite to verifying this hypothesis, we must first briefly define the commons and then see how this concept is adapted to the tensions that characterize biobank governance. We will then sketch the characteristics of genomic medicine to grasp how it is connected to biological sample collections in their present configuration.

1. According to three authors who recently devoted a dictionary to the polysemous notion (Cornu et al. 2017) of the commons, the resurgence of interest in this concept in the social sciences dates to the 2009 Nobel Prize to the economist and political scientist Elinor Ostrom and may be explained in two ways. From the perspective of social practices, goods may be shared more often, voluntarily or not (free software, shared bicycle or automobile rentals, shared gardens, etc.); from a political and legal perspective, the role of the state and of public and private property may require reinterpretation, to adapt both the legal contours of

[2]On big data, see the report of the *Nuffield Council on Bioethics*, "The collection, linking and use of data in biomedical research and health care: ethical issues", February 2015, available at http://nuffieldbioethics.org/project/biological-health-data.

appropriation and the ways we manage our resources and to confer new rights to the communities of *commoners* whom we must still identify.

2. If biobanks can be thought of as a modern commons, it is because they are viewed as a health tool with the twofold dimension in research and treatment that characterizes their development and thus as public goods, with the notable exception of the collections held by private pharmaceutical companies, which exploit them for their profit. On the other hand, they are the object of exclusivist temptations through claims of ownership of the samples and data that necessarily comprise them: institutions or researchers may consider that the collections that they have worked over time to assemble are their own, and communities of sick people may refuse access to their samples except in drastic conditions, etc. The qualification of commons may thus be more prescriptive than descriptive, intended to fulfill the function of thwarting exclusivist temptations, by ensuring that the rights claimed by some stakeholders—biobankers, donors, researchers, industries, etc.—are not exercised to the detriment of the general or public interest.

   This process of sharing has prospered in recent years, under various forms, with varying intensities, and on different legal bases: sometimes the commons are legal, such as the FAO (*Food and Agriculture Organization*) international treaty on phytogenetic resources for food and farming (2009); at other times they are those enforced by stakeholders, such as the codes of conduct by which the European Society of Human Genetics, for instance, has long battled against turning collections into private sanctuaries (Bellivier and Noiville 2009; Lenk et al. 2011). It is nonetheless a fragile process. On the one hand, it includes aspirations that are sometimes difficult to reconcile: the assembling of samples and data simultaneously to store, circulate, develop, redistribute, etc. On the other hand, public authorities have paid little attention to the question of the links between these needs, despite their support of large-scale initiatives as the European Biobank Network[3] and although biobank activities basically concern every individual who might possibly provide resources to a biobank or benefit from the results of research conducted with these resources. The absence of specific legislation (except for two articles in the Public Health Code[4]) and of public debate outside of specialized circles raises questions about the organization of biobanks and the use of their contents.[5] The questions are amplified when we look at biobanks in the light of the recent development of what we call genomic medicine.

3. In the United States, in China, Great Britain, Estonia, and the Netherlands, doctors and researchers are increasingly focusing on the whole-genome

[3]See http://www.biobanques.eu/fr/.

[4]Articles L. 1243-3 and L. 1243-4 of the Public Health Code.

[5]For an overview of the activities of biobanks in France, see Washetin et al. (2017).

sequencing of patients, intended to become a new model of health strategies.[6] This is not to say that the sequencing of the human genome is unprecedented or even rare: since the first reading in the 1990s, sequencing of patient genomes has become part of hospital screening practices and of the treatment of rare diseases and cancers in France and elsewhere. The French national cancer institute (INCa), in particular, has set up platforms to study the molecular genetics of tumors, aiming to sequence the genes and identify the mutations responsible for diseases, as well as to demonstrate the heterogeneity of tumors of the same organ, which should help lead to targeted therapies. In these situations, nonetheless, it is not the whole genome that is sequenced but groups of genes. In research, at best, the exome is sequenced, that is, the part of the genome constituted by exons, gene segments that are expressed as proteins.[7] Broadly, the objective of genomic medicine is to take samples from patients, to sequence the whole genome (which is becoming rapid and relatively inexpensive), to draw from it data that make it possible for the physician to refine the diagnosis and adapt the treatment, and then to store these data, which should ultimately constitute the substrate for new research in health and thereby improve our understanding of major diseases.

France is now turning down this road, with its *Plan France Médecine Génomique* 2025, which aspires to create the conditions under which whole-genome sequencing becomes, as soon as possible, routine. Sequencing must thus be integrated into the health care pathway according to a continuum (whence the notion of "genomic health care pathway") starting with four major sectors: metastatic or treatment-refractory cancers, rare diseases, diabetes, and then the general population, as France is late in general population sequencing, which is nonetheless useful for measuring genetic heterogeneity. By 2020, approximately 235,000 genomes should be being sequenced annually, from patients with samples taken during this early period only in university hospital centers, for indications validated by the French national authority for health.[8] On July 17, 2017, Prime Minister Edouard Philippe launched the first two sequencing platforms, SEQOiA and AURAGEN.[9]

[6]See, for Great Britain, the 100,000 Genomes Project, 2014, https://www.england.nhs.uk/publication/100000-genomes-project-paving-the-way-to-personalised-medicine/ and https://www.genomicsengland.co.uk and, for the United States, Precision Medicine Initiative, 2015, https://obamawhitehouse.archives.gov/the-press-office/2015/01/30/fact-sheet-president-obama-s-precision-medicine-initiative, and https://allofus.nih.gov/.

[7]See these two sequencing projects: the French Exome Project (FREX) and the France GenRef Project.

[8]See Aviesan, 2016, available at https://www.aviesan.fr/aviesan/accueil/toute-l-actualite/plan-france-medecine-genomique-2025. This plan grants the wish expressed, among others, by the Academies of Medicine and of Technologies in their joint opinion in favor of increased integration of ultra-high-throughput sequencing into medical practice. See Académies de médecine and Académie des technologies, *Rapport et recommandations sur la mise en œuvre en France des techniques de séquençage de nouvelle génération*, 2016.

[9]See the government press release, dated July 17, 2017, available at http://www.gouvernement.fr/partage/9344-plan-france-medecine-genomique-2025-lancement-des-2-premieres-plateformes.

These major projects will certainly affect the design and operation of biobanks, in at least three ways.

First, making sequencing systematic and integrating it into the routine health care pathway will have the mechanical effect of multiplying the production of great masses of the biological samples and data intended to fill these biobanks, which cannot help but affect their operations.[10] Then, through these programs, the biobanks, which we have been able until now to define as research infrastructures, will change their physiognomy as they move to the crossroads of individual medicine and research. The placement of sequencing at the heart of the health care pathway explains the need for access to biological resources other than those already stored in existing biobanks, which nonetheless lose none of their interest.[11] The aim now is to sequence the DNA of a given patient, above all for the purpose of providing him or her with better medical care. Even if the DNA of this specific patient was already stored in a biobank on an earlier occasion, the technical developments in terms of analysis and storage of samples mean professionals cannot be satisfied with what they have. This calls attention to what has long been one of the greatest issues of big data, in all domains: its permanent updating.[12]

The research enabled by biobanks is itself affected by the scope of the objectives: there is a clear move from biobanks for a project to infrastructures for all projects, since now researchers can probe the patient's whole genome without any initial pathophysiologic hypothesis and, finally, the genomes of the whole population, to measure its genetic heterogeneity. Finally, genomic medicine places biobanks in a more strategic perspective than ever before, because, as affirmed by each of the states mentioned above with large-scale sequencing programs, sequencing will also allow the development of a new economic and industrial sector that will include the sequencing itself (for which a US company, Illumina, currently holds a near monopoly), computer sciences (data management and computation), pharmacology, and data security.

The development of genomic medicine should probably affect the legal questions raised by the banking of human bodies. Some of the classic issues (consent, right to withdrawal, feedback of information, etc.) are revived by the systematic nature and scale of the sequencing. One such question involves the purpose of the consent: must it change from specific (for a study) to global? Another is feedback to the donor: information considered incidental is usually excluded, but the concept of incidental information is obsolete once researchers are looking in all directions for anything: if there is no specific starting hypothesis, there is no longer any information found by accident (Schröder 2017). Yet another issue is information about relatives, already

[10]In Great Britain, as part of the 100,000 Genomes Project, see https://www.genomicsengland.co.uk/taking-part/the-process/samples/.

[11]See the interview of Rory Collins, director of the UK Biobank, published by the Switzerland daily newspaper, *Le temps*, on April 28 2016, available at https://www.letemps.ch/sciences/2016/04/28/rory-collins-biobanques-acquerront-plus-plus-valeur.

[12]More information on this point can be found in the report devoted by *The Science Magazine of the Max Planck Society* to *Big Data* and, in particular, to the historical presentation by Heiborn (2017).

so complicated (who must inform the relatives of a person in whom a hereditary predisposition to any disease has been discovered? about what? and how?)[13]; it will become still more complex with the increase in the number of genomes sequenced. Finally, how can the question of data security, which is more sensitive than ever, be dealt with: are the security systems and protocols sufficient for the volume and nature of the data engendered by these projects? How should potential requests from police who want access to these data be treated? (Bellivier and Noiville 2009).

We will however put these key questions aside to turn to another that is a prerequisite to answering them. What profile does France plan to give to what will be the new generation of biobanking: open, locked, or shared? Although it is anything but unequivocal, the concept of the commons is useful in that it makes it possible to uncover the legal representations, existing or potential, desirable or objectionable, at the heart of public policies. Given that France's plan for genomic medicine for 2025 is still at its very beginnings, we will analyze these issues from the examples of more advanced foreign programs and, more particularly, the 100,000 Genomes Project underway in England since 2014.

18.2 The Commons, Tested by Practices

On examination, three findings emerge. The exclusivist rationale for the use of the resource has been rejected in favor of guaranteed access (1). The same should be true for the development of the resource through intellectual property, although vigilance on this point is appropriate (2). We are more doubtful, however, about the stakeholders in governance; for this criterion, the commons appears in some respects to serve only as a facade (3).

1. The plans for the use of genomic medicine resources restate the norm that has been largely accepted in biobank practices, that is, a policy of openness intended to combat the impasse created by the exclusivist point of view. This choice is confirmed by two points of the British and French plans.

   In the first place, the plans make no mention of any concept of physical property. No reference whatsoever is made to this category, neither for the biobank nor for the donors; a property right for donors over his or her samples and data would be dangerous for the biobanks, granting the donors prerogatives contrary to the biobanks' objectives, to sell either, for example. Accordingly, the UK Biorepository established by Genomics England to store the biological samples must "keep, store, and process" this valuable resource, but is never described as the owner. This is a remarkable shift compared with the UK Biobank, which is the "legal custodian" of the samples and the data,[14] that is, a

[13]For the French system, see Articles 1131-1 et seq. of the Public Health Code.

[14]About this qualification, see https://www.ukbiobank.ac.uk/wp-content/uploads/2011/05/EGF20082.pdf).

status resembling a form of right over the thing, even if it is a right exclusively aimed at protection. Nonetheless, these two statuses come down to the same thing, for neither institution has the prerogatives of a sole proprietor. Moreover, the difference is probably explained by the fact that Genomics England is a private company, while the UKBiobank is a charitable company. It is inconceivable that the project would have been socially accepted, had a private company been granted property rights in the data and samples.

The conditions of guaranteed access to the resources, their regrouping, and the openness of the access demonstrate the intent to share.

First, in Great Britain, the resources are centralized in premises specially planned for this purpose, thus, the samples are in a specifically designed UK Biorepository in Milton Keynes. In principle, that is not the direction France is taking; for once, it has chosen a decentralized system, with 12 platforms and housing the samples in the existing biobanks,[15] at the same time as the data are brought together in a central analysis service (*Collecteur analyseur des données*, CAD).[16] But regardless of the type of infrastructure chosen, at a minimum the biobanks are networked, harmonizing their practices to ensure their smooth and secure circulation. This double procedure is intended to use the samples and data to the greatest advantage possible and to ensure better data security.

Next, an environment propitious to research has been set up in the UK *via* the GeCIP (Genomics England Clinical Interpretation Partnership), open to private and public participants, on condition, however, that the project be both scientifically serious and ethical, as determined by the Genomics England data access committee.[17] The users can also request a period of exclusive access to the data but for only 6 months; afterwards the data are again made available to the community of researchers.

2. Although the idea of commons does not at all exclude the development of resources, including by intellectual property rights, the value drawn from them must not benefit only a few to the detriment of the community (Bellivier et al. 2015, p. 197). This philosophy underlies the policy statement of Genomics England: "all GeCIP users will be required to contribute results and data to the GE data center to enhance the scientific knowledge base". They thus have a maximum of 9 months to publish their results.[18]

[15]France is likely to rely on the already existing platforms (at the Curie Institute, INCa, etc.) for the storage of samples.

[16]About the CAD, see the report *France Genomique*, 2015, p. 22 et seq., available at http://www.gouvernement.fr/sites/default/files/document/document/2016/06/22.06.2016_remise_du_rapport_dyves_levy_-_france_medecine_genomique_2025.pdf.

[17]For its composition, see https://www.genomicsengland.co.uk/about-genomics-england/the-board/access-review-committee/.

[18]*Genomics England Publication Policy*, 2016, available at https://public.huddle.com/a/pZZBloE/index.html.

A similar creed infuses its intellectual property policy.[19] First, Genomics England agrees that it will not apply for any preemptive patents, for example, of isolated gene sequences, or that are overly broad, to avoid the problems for researchers presented by excessively extensive protection of these basic research tools. Next, the objective is for Genomics England to own and control as far as possible the intellectual property (jointly with third parties who have participated in inventions through GeCIP), which is not only practical for companies who can thus have access to a single pool but also constitutes a guarantee about how the patents are used. Accordingly, Genomics England has stated its preference for granting non-exclusive licenses. Finally, when a patent for a medical product or device patent is to be developed and produced, Genomics England will endeavor to negotiate a preferential price for the National Health Service.[20] This policy is inspired by that of the UK Biobank: researchers who obtain results can patent them according to intellectual property rules, but in return they must grant a free license to the UK Biobank that allows it to make these results available to other researchers.[21]

But we will pay attention to how these rules, which clearly pursue the common interest that Genomics England seeks to enforce, will be implemented in a legal, administrative, and economic environment tugged back and forth for 30 years between exclusivism and openness. That is, they should not remain meaningless declarations of principle.

3. Examining the governance planned in these genomic medicine programs, however, we are troubled by some aspects of the distribution and the detailed coordination of the roles attributed to the stakeholders of the commons. On the one hand, although the role of the state is contested in most situations of commons, here it is primary. On the other hand, although a commons is normally based above all on the stakeholders themselves, here donors have very little power—probably a poor choice.

   (a) First, the idea that the state is the guarantor of the commons is generally strongly criticized, to the point that some argue that the commons cannot prosper within the logic of the state, but is rather a third term between the state and the market. In this domain, however, the state remains central. This is one

[19] *Genomics England Intellectual Property Policy*, 2017, available at https://www.genomicsengland.co.uk/about-gecip/for-gecip-members/documents/.

[20] The objective is to avoid the difficulties raised in Europe by the patents protecting the BRCA genes. On this point, see D. Stoppa-Lyonnet et M. Cassier *in* F. Bellivier et Ch. Noiville, *La bioéquité, Batailles autour du partage du vivant, Autrement, coll. Frontières, Paris, 2009*, pp. 21 et seq., pp. 34 et seq. The question must be put into context with the more general one of drug prices and attempts to reach a "fair and controlled" price, in the words of the French Myopathy Association (see http://www.usinenouvelle.com/article/pour-l-afm-telethon-la-derniere-etape-est-de-mettre-les-traitements-a-disposition-des-patients.N471603), especially in today's drug price explosion, especially for cancer therapy (see https://curie.fr/actualite/ouverture/observatoire-cancer-cout-des-traitements).

[21] See http://www.ukbiobank.ac.uk/wp-content/uploads/2011/11/Access_Procedures_Nov_2011.pdf, point B 8.

of the key points of the French plan: while private companies are necessarily being approached for some aspects of its implementation (sequencing, data analysis, etc.) and although the plan should result in numerous public-private partnerships, it is under the leadership of the state. In Great Britain, the Ministry of Health created Genomics England, a limited liability company, to implement the 100,000 Genomes Plan, but also worked in close collaboration with diverse public institutions, including the NHS, with whose transformations it must be in phase.[22] The state is thus playing the role of guardian of the resources; although private companies certainly participate in various capacities (governance, subcontracting), the most sensitive aspects, such as management of the sequencing data (its storage and circulation) are in the hands of the public stakeholders. Although this role is delegated to private participants in the United States and China, with the risks of development contrary to patients' interests that can result, Europe appears to be moving in the opposite direction.

(b) Second, given that they emanate in the strict physical sense of the word from people-resources who have made available material from their own body and their personal data, we must ask whether they can bypass these essential stakeholders in their governance. In the background of the biobanks, emerging communities consider themselves the legitimate recipients of the fruits of these resources: communities of researchers, of physicians, but also of donors. For the donors, at least, the hypothesis of the commons remains to be demonstrated, although giving patients confidence by involving them is a *sine qua non* of this type of company, especially given that for these "ordinary humans" the sequencing of the human genome is at the least a complex business and the need to collect great masses of data to improve health care not at all obvious.

Certainly, in one way, donors are central: their rights are carefully reaffirmed, including the need for informed consent, the particular situation of vulnerable people, the right to withdraw consent, and to feedback about the study's results (individual or collective, depending on the legal system). Everything is organized so that donors consent on a fully informed basis to their treatment and to the research. As indicated by the informed consent forms in Great Britain—and clearly those under preparation in France as well, nothing about the possibilities of development and commercialization is hidden from them, although numerous questions remain

[22]"The Department of Health has established Genomics England as a wholly owned, limited company to deliver the project. Genomics England is working with NHS England (NHSE), Public Health England (PHE), Health Education England (HEE), NHS Trusts, the Northern Ireland Department of Health (DoH NI) and a number of HSCNI organisations". See The 100,000 Genomes Project Protocol, January 2017, p. 6, available at: https://www.genomicsengland.co.uk/wp-content/uploads/2017/03/Genomics-England-Protocol-v3.pdf.

open.[23] Nor are the ethical questions about the donors evaded, far from it; in Great Britain, they are debated by an Ethics Advisory Committee, whose task is to ensure that the 100,000 Genomes Project is conducted in the interests of all its stakeholders. More generally, the *Plan France Médecine Génomique 2025* illustrates a public policy that builds the genome sequencing onto the French concept of health as a public good, precisely so that donors are not forgotten. It therefore qualifies sequencing as medical and thus covered by the national health insurance fund, which does raise two questions.

First, in this period when we seek to control healthcare costs, what will be the impact—and on whom—of this vast operation?

Finally, knowing that in France we long ago—rightly—abandoned the idea of compensating donors in cash, or even strictly reciprocally (someone who makes available his or her own physical resources and personal data will not necessarily receive treatment based on results in return), what form of recompense, albeit symbolic, can we imagine? Acknowledgements in publications? Fair drug prices? Financial contribution to the storage of the physical resources and data from those who obtained a patent due to their own work, but also due to the participation of the donors? (Bellivier 2017).

These vast plans for genomic medicine thus revive considerably several fundamental questions, including those at the shifting frontier between care and research and those of the interest of the accumulation and organization of immense masses of physical resources and personal data. Before it is too late to turn back, the concept of the commons could be a good angle of attack to understand what is at work in the plans currently under development. It is not enough to say that everyone is potentially concerned by the disease. What is necessary is to find the means to organize their voluntary and responsible involvement.

References

Bellivier, F. (2017). Biobanques. In M. Cornu et al. (Dir), *Dictionnaire des biens communs* (pp 127 et seq). Paris: PUF.

Bellivier, F., Benhamou, F., Cornu, M., & Noiville, C. (2015). Collections muséales et collections biologiques : de la conservation à l'accès? In B. Coriat (Ed.), *Le retour des communs. La crise de l'idéologie propriétaire, propriétaire* (pp. 197 et seq). Paris: Editions Les liens qui libèrent.

Bellivier, F., & Noiville, C. (2009). *Les biobanques*, Que Sais-Je ?, PUF, 2009 and L*a bioéquité. Batailles autour du partage du vivant*, Autrement, Coll. Frontières.

[23]See, for example, the questions raised by the Science and Technology Committee, House of Commons, "Genomics and genome-editing: future lines of inquiry", Sixteenth Report of Session 2016–17, May 1, 2017: "[t]he suitability of the 'broad consent' model being employed by the 100,000 Genomes Project, and whether the consent materials and patient recruitment techniques fully inform participants of the potential commercial uses of their data. This is tied to the debate around the nature of genetic data and whether it deserves a privileged status over other forms of personal health data".

Bellivier, F., & Noiville, C. (2018). Les collections d'échantillons biologiques à l'heure de la médecine génomique: un nouvel exemple de bien commun? In X. Bioy (Dir), *La régulation publique des centres de ressources biologiques. Le cas des tumorothèques* (pp 391 et seq). Les Etudes hospitalières.

Boldin, B. (2015). Economie de la santé comme bien public mondial. *Revue de la régulation* [Online], 17 | 1st semester, Spring 2015, posted June 17, 2015 at http://regulation.revues.org/11128

Chamayou, G. (2008). *Les corps vils. Expérimenter sur les êtres humains aux XVIIIème et XIXème siècles.* La découverte.

Cornu, M., Orsi, F., & Rochfeld, J. (Eds.). (2017). *Dictionnaire des biens communs*, PUF, collection Quadrige.

Heiborn, T. (2017). Stacking data. *The Science Magazine of the Max Planck Society, 2*, 27 et seq. https://www.mpg.de/11357066/MPR_2017_2.pdf

Lenk, C., Sándor, J., & Gordijn, B. (2011). *Biobanks and tissue research. The public, the patient and the regulation*. Heidelberg: Springer.

Mills, P. (2016). For whom the bell tolls: Precision medicine, private virtue and the public good. http://nuffieldbioethics.org/blog/for-whom-the-bell-tolls-precision-medicine-private

Schröder T. (2017). Treasure hunt in the data jungle. *The Science Magazine of the Max Planck Society, 2*, p. 34 et seq. https://www.mpg.de/11357066/MPR_2017_2.pdf

Washetin, K., Ilié, M., Hofman, V., Butori, C., Long, E., Lassalle, S., et al. (2017). Les biobanques : quels enjeux en 2017 ? *Revue francophone des laboratoires, 2017*(488), 25–29. http://www.sciencedirect.com/science/article/pii/S1773035X16304166?via%3Dihub.

Erratum to: Public Regulation of Tumor Banks

Xavier Bioy

Erratum to:
X. Bioy (ed.), *Public Regulation of Tumor Banks*, https://doi.org/10.1007/978-3-319-90563-1

This online version of this book was inadvertently published without a Copyright remark about the French version in the copyright page. This has now been rectified.

The updated online version of this book can be found at
https://doi.org/10.1007/978-3-319-90563-1

X. Bioy (ed.), *Public Regulation of Tumor Banks*,
https://doi.org/10.1007/978-3-319-90563-1_19

The manufacturer's authorised representative in the EU is Springer Nature Customer Service Centre GmbH, Europaplatz 3, 69115 Heidelberg, Germany. If you have any concerns regarding our products, please contact ProductSafety@springernature.com

Printed and bound by CPI Group (UK) Ltd, Croydon, CR0 4YY
05/10/2026
02228138-0007